Physical Geography in Diagrams

Metric Edition

R. B. Bunnett, B.Sc. (London)

Longman

LONGMAN GROUP LIMITED
Longman House,
Burnt Mill, Harlow, Essex.

*First published *1965*
*Second Edition *1973*
*Third Edition *1976*
Fifth impression 1981

ISBN 0 582 34122 1

Acknowledgements

For permission to reproduce the photographs on the pages mentioned below,
the Publishers are indebted to the following:
Aerofilms, 71 right, 82 top, 170; J. Allen Cash, 49 bottom, 53, 57, 65 bottom, 82
bottom; 116, 117 top left, 167, 171; American Museum of Natural History 87;
Ansel Adams, Magnum 175; Australian News & Information Bureau, 69, 168
top; 174 left, Barnaby's 168 bottom, 174 right, H.H. Bennett, F.A.O., 180 centre;
Victor Bianchi, 182 top; B.O.A.C. 19 top, British Information Service, 65 top,
R.B. Bunnett 19 bottom, 21, 44 right lower, 76 left, 78, 79; Camera Press 182
bottom; H.E. Dale, U.S. Department of Commerce Weather Bureau, 137;
Department of Highways, South Dakota, 44 bottom left; Ewing Galloway 57
bottom right; Exclusive News Agency 70; Fairchild Aerial Surveys 59; French
Government Tourist Office, 35 bottom left; Geological Survey Photograph 76
right, 84; R. Hertzog, 68; Hong Kong Government Information Service
182 centre; W. Heerbrogg 35 top; I.P.S. (U.S.I.S.) 31, 51 centre; Institute
Geographique National 96; J.K. Joseph 94 bottom; J. Launois, Black Star, 54;
Middle East Archives 40; Mustograph 117; Paul Popper 29, 71 top, 97, 172, 180
bottom, Quebec Government 77; Radio Times 2 (from Hale Observatory), 34 top
right, 72 left; Sky Foto 34 left, S.A.S. 82 top; Societe Encyclopedique Universelle
180 top; Society for Cultural Relations with U.S.S.R. 183 bottom; Swissair
Photo 93 bottom; Swiss National Tourist Office 44 top, 91, 93 top, 94 top; U.S.
Forest Service 173, U.S.I.S. 95, 181, 183 top, 184 top and bottom, 185 top and
bottom; Zambia Information Dept. 51 left; The publishers also wish to thank
the Cambridge Local Examination Syndicate for permission to reproduce past
examination questions.
   The publishers have made every endeavour to contact the holders of copyright
to all the photographs used but in some cases without success and to these the
publishers offer their apologies.

Printed in Singapore by Selector Printing Co. Pte. Ltd.

Preface

In spite of recent changes in the teaching of geography up to
the General Certificate of Education 'O' Level, many teachers
acknowledge that physical geography remains a prerequisite for
human geography. This book enables students to make a thorough
study of the processes which operate to produce man's environment
and the physical features these produce.

In this third edition the author brings the theories and concepts
related to the origin of physical features, into line with current
thinking. The Theory of Plate Tectonics is used as a possible means
of explaining the formation of continental masses and major features
such as fold mountains, plateaues, rift valleys and ocean deeps, etc.
A large part of the chapter dealing with the work of running water
has been re-written, and additional materials have been included
in the chapters on weather and climate.

This book is intended to be used by students during the last two
years of the geography course leading to the General Certificate of
'Education 'O' Level, and other similar examinations. It covers all
the material required in answering all types of questions set on
physical geography in these examinations. Diagrams are used
throughout the book as the medium through which the subject
is approached because the author believes that large, well-annotated
diagrams are the best way of getting most students to understand
easily, geographical processes and the features they produce. The
book contains over 500 line diagrams and more than 70 photographs,
and wherever possible these are used to explain geographical
principles and concepts. The text is superimposed on this foundation.

Since this book is used by students entering for 'O' Level
Examinations, a set of exercises has been given at the end of each
chapter. In addition, further revision exercises are given at the end
of the book together with questions from recent examination papers
set by the various Examination Syndicates. The author has also
included multiple choice of questions, with both four and five
responses, at the end of each chapter, because he believes that
these are of considerable value in assessing a student's understanding
of the basic principles and concepts of physical geography. These
questions can also be used to identify some of the misconceptions
held by, and various aspects of physical geography that present
difficulties to, some students.

R.B.B.
1976

Objective Questions

In recent years, several countries have replaced traditional questions with objective questions in the Certificate 'O' Level Geography Examinations, and it is likely that many more Examining Authorities will follow suit.

It is important that students obtain training and experience in understanding and correctly answering objective questions, and because of this objective questions have been set on all the work in this book. These are given at the end of each chapter and in addition there are Two Tests, each covering all aspects of Physical Geography. These are at the end of the book. The traditional type of questions are retained because many teachers and students still require these.

Objective questions can be set with either four or five answers to choose from, and because some Examining Authorities set questions with four answers, and others set questions with five answers, it has been decided to give both types of question in this book.

Objective questions are of several types – some are factual and some are deductive, and some are easier than others. This book contains three types of objective questions:
(1) the *'completion'* type   (2) the *'question'* type, and
(3) the *'negative'* type.
1. In the *'completion'* type, a student is asked to select the correct answer which when added to the question will turn it into a complete and correct statement.
    Example: Question 2, Chapter 13.
    Minimum and maximum temperatures are obtained from an instrument called
    A a barometer
    B a Six's thermometer '
    C an anemometer
    D a clinical thermometer
    Of these four answers, two (A and C) are not instruments for measuring temperature, and one (D) is used for measuring a person's temperature. Only B is correct because it is a thermometer for measuring maximum and minimum temperatures. The correct answer B is therefore shown by shading the box below B as given here.

2. In the *'question'* type, the student has to select the completely correct answer and to indicate this by shading the appropriate box. In this type of question, one or more answers may be **partly** correct, but only one will be fully correct.
    Example: Question 1. Chapter 3.
    Which one of the following groups of terms is applicable to some parts of the ocean floor?
    A basin, deep, cirque, plateau
    B trench, ridge, basin, plateau
    C plateau, basin, dune, ridge
    D ridge, deep, basin, waterfall
    Only answer B is completely correct. Each of the other answers contain three correct features and one incorrect feature. The latter are **cirque, dune** and **waterfall,** all of which are features which only occur on land surfaces. The correct answer is shown as follows.

3. In the *'negative'* type of question, which may be of the *'completion'* type or the *'question'* type, only one of the answers given is **not** true, and it is this answer that the student has to recognise.
    Example: Question 2. Chapter 3.
    Which of the following statements is **not** true in respect of sedimentary rocks?
    A the particles of rock are sometimes completely of organic origin
    B the rocks are non-crystalline
    C they are rocks whose structure is determined by great pressure or heat
    D the rocks have been deposited in layers.
Answers A, C, and D are true of sedimentary rocks, but answer B is not true because many sedimentary rocks are composed of mineral particles which are crystalline, for example, sandstone. The answer which is not true is B and this is indicated by shading the appropriate box as shown below.

Instructions for Answering Objective Questions

1. Carefully read each question to make sure that you understand what the question is asking.
2. Answer the questions in any order.
3. Do not guess the answers.
4. Each of the two tests at the end of the book contains 32 questions. You should allow yourself 40 minutes to answer each Test.
5. Before starting any test make sure that you have a piece of paper, for rough work, a ruler, an eraser (rubber) and a protractor.

Metric Conversion

LENGTH
1 metre (m) = 100 centimetres (cm)
= 1000 millimetres (mm)
1 kilometre (km) = 1000 metres (m)
1 international nautical mile = 1852 metres (m)

From metric

1 millimetre (mm) = 0·03937 inch
1 metre (m)   = 1·094 yards
= 3·281 feet
1 kilometre (km) = 0·621 mile

To metric

1 inch   = 25·4 millimetres (mm)
1 foot   = 304.8 millimetres (mm)
= 0·3048 metres (m)
1 yard   = 0·9144 metre (m)
1 mile   = 1·609 kilometres (km)

AREA
1 hectare (ha)   = 10 000 square metres (m²)
1 square kilometre (km²) = 1 000 000 square metres (m²)
1 square kilometre (km²) = 100 hectares (ha)

From metric

1 square metre (m²) = 1·196 square yards
= 10·76 square feet
1 square kilometre (km²) = 247·105 acres
= 0·386 square mile
1 hectare (ha) = 2·471 acres

To metric

1 square foot   = 0·0929 square metre (m²)
1 square yard   = 0·836 square metre (m²)
1 acre   = 4046.86 square metres (m²)
= 0·4047 hectare (ha)
1 square mile   = 2·590 square kilometres (km²)
= 259 hectares (ha)

TEMPERATURE
From °C
$t = \frac{9}{5}\theta + 32$
Where θ = temperature in degrees Celsius (Centigrade) (°C)
t = temperature in degrees Fahrenheit (°F)

To °C
$\theta = \frac{5}{9}(t - 32)$
Where θ = temperature in degrees Celsius (°C)
t = temperature in degrees Fahrenheit (°F)

For a more complete list of conversions see 'Metric and Other Conversion Tables' by G.E.D. Lewis (Longman 1973).

Contents

1   Our Home in the Universe

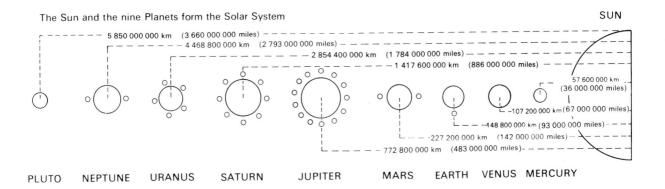

The Sun and the nine Planets form the Solar System

SUN

5 850 000 000 km   (3 660 000 000 miles)
4 468 800 000 km   (2 793 000 000 miles)
2 854 400 000 km   (1 784 000 000 miles)
1 417 600 000 km   (886 000 000 miles)

57 600 000 km
(36 000 000 miles)

107 200 000 km (67 000 000 miles)

148 800 000 km (93 000 000 miles)

227 200 000 km   (142 000 000 miles)

772 800 000 km   (483 000 000 miles)

PLUTO   NEPTUNE   URANUS   SATURN   JUPITER   MARS   EARTH   VENUS   MERCURY

INTRODUCTION TO THE UNIVERSE

At night the sky often appears to be full of stars each of which seems to be no bigger than a twinkling speck. But it comes as a surprise to learn that every star is much bigger than the earth: indeed some are several millions of times bigger. Again, the distances between stars in the night sky do not appear to be very great, but astronomers have calculated that despite the millions of stars in the universe, they are so scattered in space that together they occupy only a very small part of space. Some idea of the immensity of the universe can be obtained by considering speed, distance and size. Light and radio waves travel 298 000 kilometres or 186 000 miles per second which means that if we transmitted a radio signal from earth it would take about $1\frac{1}{4}$ seconds to reach the moon, 8 minutes to reach the sun and almost 4 years to reach Proxima Centauri, our nearest star, and just under 20 years to reach Delta Pavonis, a slightly more distant star. Stars tend to form clusters, which are known as *galaxies*, and galaxies form *groups*. Our Local Group, that in which the earth is located, contains 27 galaxies. In this Local Group is a distant galaxy, just visible with the naked eye, which is called Andromeda Spiral. A radio wave transmitted from earth would take about 2·2 million years to reach this galaxy. In other words, when we look at Andromeda Spiral we are seeing it as it was 2·2 million years ago. Distances are so immense that it is impossible for us to draw a scaled map of the universe. Let us try doing this by using an infinitesimally small scale. On this scale let us regard the sun as being the size of a hydrogen atom, that is, one 100 millionth of a centimetre in diameter. Our galaxy would then have a diameter of 69 metres (225 feet) and the sun would be about 25 metres (82 feet) from its centre. Andromeda Spiral would be about 1·6 kilometres (1 mile) away and the edge of the known universe would be almost 3200 kilometres (2000 miles) away.

THE SOLAR SYSTEM

This system contains the sun and its nine planets which revolve around the sun in elliptical orbits. The light of the sun falls on each of the planets and it is in turn reflected by them. This is the only light that the planets reflect.

All the energy of the Solar System is derived from the sun whose surface is covered with burning gases and whose temperature is about 6000°C. Mercury, which is the smallest planet, is nearest to the sun while Pluto, which is smaller than Earth, is the farthest away from the sun. Some of the planets, eg. Earth, Jupiter and Saturn have *satellites*: the moon is the satellite of the Earth.

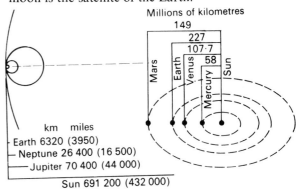

Millions of kilometres
149
227
107·7
58

Mars   Earth   Venus   Mercury   Sun

km   miles
Earth 6320 (3950)
Neptune 26 400 (16 500)
Jupiter 70 400 (44 000)
Sun 691 200 (432 000)

The Andromeda Spiral Nebula

Because the planets are at varying distances from the sun, and because they revolve around the sun, they each take a different time in which to complete one orbit. Mercury completes its orbit in 88 days, that is, a year on Mercury lasts 88 days. The earth completes its orbit in $365\frac{1}{4}$ days, which is the length of one year on earth. The moon, which revolves around the earth, takes approximately 27 days to do so.

PHASES OF THE MOON

As the moon revolves around the earth the illuminated part of it apparently varies in size. In the diagram the two circles represent moon positions. The outer circle clearly shows that exactly half the moon is illuminated all the time. The inner circle shows what the moon looks like to us on earth during its different positions. The illuminated part of the moon takes different shapes – at full moon it is a circle. Look at the moon on different nights in any one month and find out whether that part of the moon which is not illuminated can be seen.

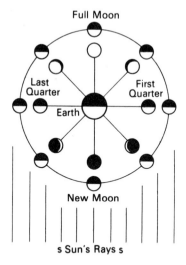

Eclipse of the Moon
(earth comes between moon and sun)

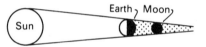

Eclipse of the Sun
(moon comes between earth and sun)

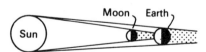

Eclipse of the Sun as seen from the Earth

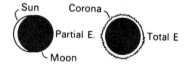

THE SHAPE OF THE EARTH

At one time the earth was thought to be flat, but today it is a proven fact that it is round. This is accepted by all. There is much evidence to show that the earth is round but before looking at this it is as well to remember that it is not quite a perfect sphere. The polar circumference is about 130 kilometres (80 miles) shorter than the equatorial circumference and the polar diameter is almost 40 kilometres (25 miles) shorter than the equatorial diameter. The earth is, therefore, slightly flattened at the poles and its spherical shape is called a *geoid*.

The evidence

1 Aerial photographs. Numerous photographs have been taken by satellites at great distances from the earth, and all of these show that the earth is spherical.

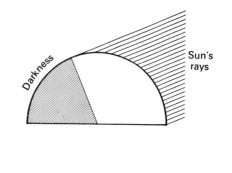

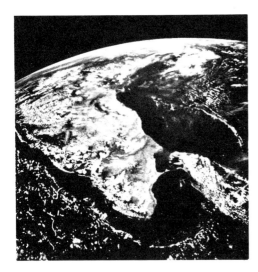

Satellite photograph showing Southern India and Sri Lanka

2 The moon's eclipse. When there is an eclipse of the moon, the shadow of the earth which is thrown on the moon is always round. Again, only a sphere can cast a shadow which is circular.

3 Circumnavigation of the earth. The earth has been circumnavigated innumerable times by land, sea and air.

4 Sunrise and sunset. The earth rotates from west to east which means that places in the east see the sun before places in the west. This is a proven fact. If the earth was flat, all places would see the sun at the same time.

5 The earth's curved horizon. The earth's horizon, when seen from a ship, a plane, or a high cliff appears curved. The curved horizon widens as the observer's altitude increases until it becomes circular. This is how astronauts see the horizon from their space ships. If the earth was not spherical, there would be no circular horizon.

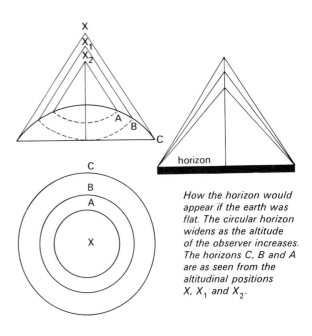

How the horizon would appear if the earth was flat. The circular horizon widens as the altitude of the observer increases. The horizons C, B and A are as seen from the altitudinal positions X, X_1 and X_2.

6 A ship's visibility. The diagram on page 4 shows two ships but you can see that the observer is only able to see one of them. If the earth was flat, the observer would see both ships.

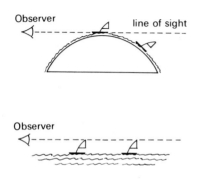

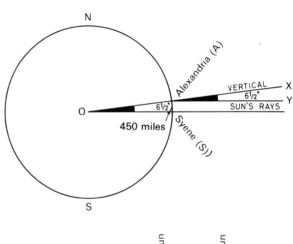

THE SIZE OF THE EARTH

About 200 B.C. Eratosthenes measured the altitude of the mid-day sun at Alexandria and found that it was $83\frac{1}{2}°$. He knew that at the same time on this particular day the altitude of the sun at Syene, which was about 720 kilometres (450 miles) to the south, was 90°. Eratosthenes knew that angle XAY was $6\frac{1}{2}°$ and that angle AOS was also $6\frac{1}{2}°$. He then calculated that if an angle of $6\frac{1}{2}°$ is subtended by an arc of 720 kilometres (450 miles) then an angle of 360° would be subtended by an arc of 40 000 kilometres (25 000 miles).

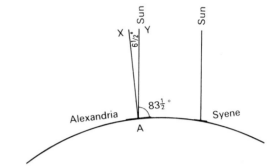

THE POSITION OF A PLACE ON THE EARTH'S SURFACE

Let us take a large ball and mark two points on it so that they are exactly opposite to each other. Now draw a line right round the ball so that it is midway between the points all the way. The line will now divide the ball into two equal parts, and, because the ball is a sphere, each part can be called a *hemisphere*. We will now call the line the *equator*

and you will see that it is a *circle*. One point we will call the *North Pole* and the other the *South Pole* (*fig. a*).

We can now draw more circles parallel to, and to the north and south of the equator. These can be called *parallels* or lines of *latitude* (latitude referring to the angular distance north or south of the equator). This idea is applied to the earth. The equator is given a value of 0°, and, as you can see

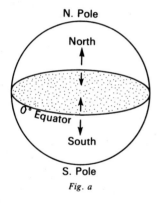

Fig. a

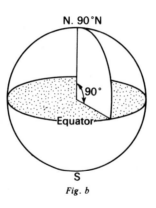

Fig. b

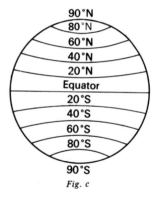

Fig. c

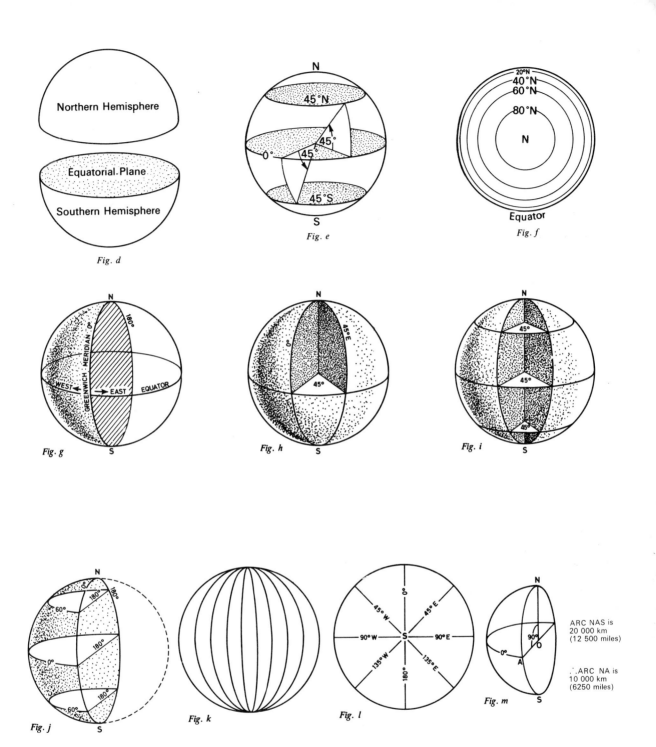

Fig. d

Fig. e

Fig. f

Fig. g

Fig. h

Fig. i

Fig. j

Fig. k

Fig. l

Fig. m

ARC NAS is
20 000 km
(12 500 miles)

∴ARC NA is
10 000 km
(6250 miles)

from *fig. b*, the North Pole has a latitude of 90°N. The South Pole has a value of 90°S. and every other place on the earth's surface has a latitude of so many degrees north or south (of the equator). You will also notice that the equator is the longest parallel. *Figs. c* and *f* show what the parallels look like on a globe from the side and from the North Pole respectively.

We can now draw on the ball another set of circles all of which pass through the two poles (*fig. k*). That part of each circle between the poles can be called a *meridian* or line of *longitude*. This idea is also applied to the earth and the meridian which passes through Greenwich is given a value of 0°; the meridian that is opposite to it will therefore have a value of 180° (*fig. g*). *Longitude refers to the*

5

angular distance east or west of the Greenwich Meridian, and all places except those on meridian 180° will therefore have a longitude of so many degrees east or west (of Greenwich). *Figs. k* and *l* show what the meridians look like from the side and from the South Pole respectively.

How long is 1° of Latitude?

Fig. m is a diagram of a hemisphere and N and S stand for the North and South Poles. Angle NOA is 90° and this is the latitude of N or the angular distance of N from the equator (0°). This angle is subtended by arc NA whose length is one half of a meridian. On the earth arc NA has a length of 10 000 kilometres approx.

If an arc of 10 000 kilometres subtends 90° then an arc of $\frac{10\,000}{90}$ kilometres subtends 1° i.e. *1° of latitude represents 111 kilometres approx.*

How long is 1° of Longitude?

Every parallel has an angle of 360° at its centre and every half-parallel is subtended by an angle of 180° (*fig. j*). If the length of the parallel or half-parallel is known then the length of the arc subtended by 1° can be calculated. For the equator this is 111 kilometres, but for other parallels it is less than this because parallels decrease in size away from the equator. *1° of longitude represents 111 kilometres along the equator.*

Great Circles

Any circle which divides a globe into hemispheres is a *great circle.* The equator is a great circle and Greenwich Meridian together with Meridian 180° make another great circle. Likewise Meridian 10°E. and 170°W., and 20°E. and 160°W., make two more great circles. The number of great circles is limitless. Great circles can extend in any direction: east to west, north to south, north-east to south-west, and so on. Great circles are of equal length.

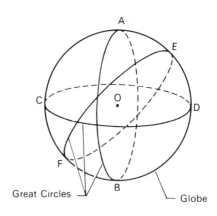

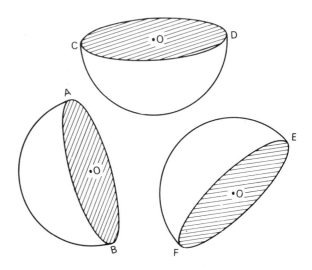

MOVEMENTS OF THE EARTH

I It rotates   II It revolves

Rotation of the Earth

The earth rotates once in 24 hours and this results in:
 (i) Day and night
 (ii) A difference of 1 hour between two meridians 15° apart

(iii) The deflection of winds and ocean currents
(iv) The daily rising and falling of the tides.

Day and Night

These four diagrams show what is happening along Greenwich Meridian during one rotation of the earth on March 21st.

The sun is rising along Greenwich Meridian. People here see it 'rising' over the Eastern Horizon.

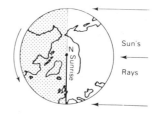

The earth has turned through $\frac{1}{4}$ of a rotation and it is noon along the Meridian. The sun has reached its highest position in the sky.

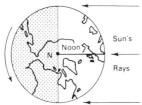

The earth has passed through $\frac{3}{4}$ of a rotation and it is midnight along Greenwich Meridian.

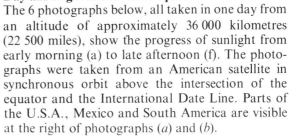

The earth has now turned through $\frac{1}{2}$ of a rotation and the sun is setting along the Meridian. People here see the sun 'sinking' below the Western Horizon.

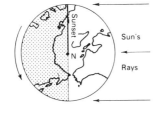

Day and Night as seen from Space

The 6 photographs below, all taken in one day from an altitude of approximately 36 000 kilometres (22 500 miles), show the progress of sunlight from early morning (a) to late afternoon (f). The photographs were taken from an American satellite in synchronous orbit above the intersection of the equator and the International Date Line. Parts of the U.S.A., Mexico and South America are visible at the right of photographs (a) and (b).

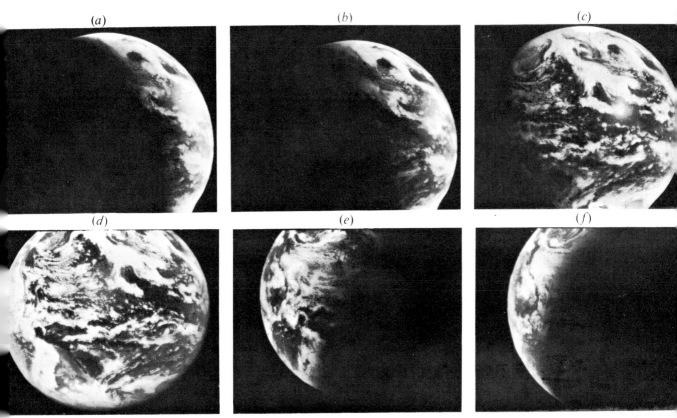

(a)   (b)   (c)

(d)   (e)   (f)

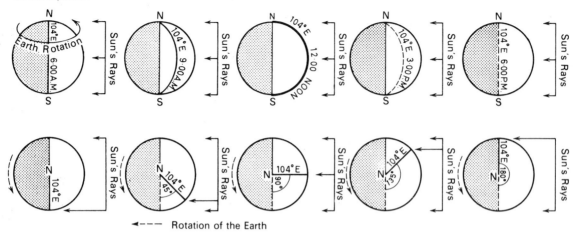

At the Equinox

◄--- Rotation of the Earth

Rotation and time

The diagram above shows the positions of meridian 104°E. at intervals of 3 hours. The top row of diagrams shows the appearance of the meridian from above the equator and the bottom row of diagrams the appearance from above the North Pole. The sun reaches its highest position in the sky for this meridian when it lies under the sun. At this time it is said to be *1200 noon Local Time* along this meridian. Local time is sometimes called *Sun Time*. The highest position of the sun for any place can be observed from a study of the lengths of the shadows cast by a vertical stick. The shortest of these is cast by the sun when it is in its highest position in the sky (study the Sun Path Diagram for Singapore).

The diagram above also shows that all places on meridian 104°E. have noon at the same time. This means that *all the places on the same meridian will have the same local time.*

Sun Path Diagram for Singapore on June 21

The vertical stick indicates the position of Singapore. The shortest shadow points due south and occurs at *noon*, i.e. when the sun reaches the highest point in its 'path' across the sky.

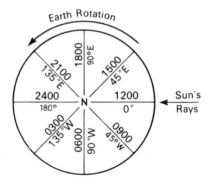

Local times for selected meridians when it is 1200 noon G.M.T.

When Greenwich Meridian lies under the sun the local time along this meridian is of course 1200 noon, but this local time is 1200 noon *Greenwich Mean Time* or G.M.T.

Behind and Ahead of G.M.T.

All meridians to the east of Greenwich Meridian have sunrise before that meridian. Local times along these meridians are therefore *ahead* of G.M.T. Meridians to the west of Greenwich Meridian have sunrise after this meridian and therefore their local times are *behind* G.M.T.

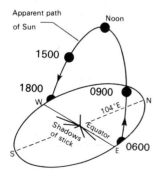

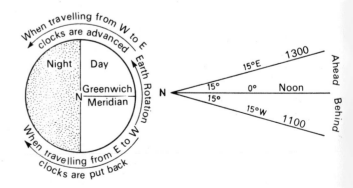

Longitude can be calculated from Local Time and G.M.T.

The local time at X is 1600 and G.M.T. is 1400. The difference in time between X and Greenwich is therefore 2 hours. This represents a difference of 30° of longitude between the two places (15° of longitude represents 1 hour). Since the local time at X is *ahead* of that at Greenwich then X is *east* of Greenwich. *The longitude of X is 30°E.* Similarly, if the local time at Y is 0800 and G.M.T. is 1400, then Y is 6 hours *behind* Greenwich, that is, Y is 90° to the *west* of Greenwich and its longitude is 90°W. *Note* If any two of the above three facts are given, the third can always be calculated.

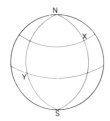

The Significance of the International Date Line

The diagram on the right shows what happens when two travellers set off at the same time (1600) on a Monday from a place A (long. 0°). One traveller goes westwards and the other eastwards to a place B (long. 180°). The traveller going west calculates the local times at 90°W. and 180° to be 1000 Monday and 0400 Monday respectively. The traveller going east calculates the local times at 90°E. and 180° to be 2200 Monday and 0400 Tuesday respectively.

In theory along meridian 180° it is both 0400 Monday and 0400 Tuesday. When the traveller going west crosses this meridian he finds it is 0400 *Tuesday*, i.e. he has *lost one day*. When the traveller going east crosses this meridian he finds it is 0400 *Monday*, i.e. he has *gained one day*. The line at which a day is lost or gained is called the *International Date Line*. This line follows meridian 180° except where this crosses land surfaces. To avoid confusion to the peoples of these regions the line bends round them so passing over a sea surface.

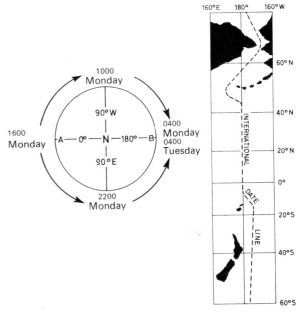

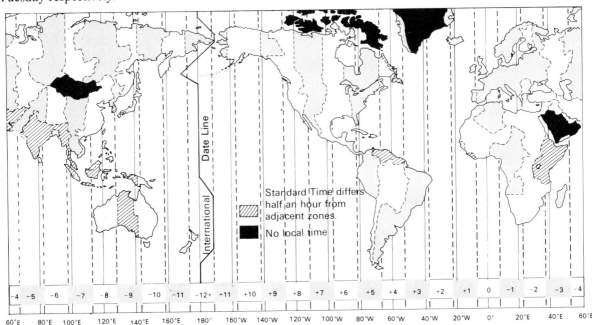

Standard Time and Time Zones

Each meridian has its own local time. Thus when it is 1200 noon local time in Georgetown (Penang) which is 100° 20′E. it is 1214 local time in Singapore whose longitude is 103° 50′E. Great confusion would arise if all places used local time. To avoid this the world is divided into 24 belts, each 15° of longitude wide. The local time of the central meridian of each belt is applied to that belt which is called a *time zone*. The local time of the central meridian for each time zone is called *standard time*. Neighbouring time zones have a difference of 1 hour. The boundaries of time zones are frequently adjusted to conform to political boundaries. (Diagram on page 9, bottom).

Deflection of Winds and Ocean Currents

All places on the earth's surface make 1 rotation in 24 hours. A place on the equator moves eastwards at a greater velocity than a place on say parallel 60°N. because the equator is longer than this parallel. A mass of air or water on parallel 60°N. will have an eastward speed equal to that of the parallel. If this mass moves towards the equator it will cross over parallels whose eastward speeds increase with decreasing latitude. The path of the mass when plotted appears as a curve which bends to the *right* (from the starting point) of the path it would have taken if the earth had not been rotating. If a similar air or water mass had moved from a high latitude to a low latitude in the Southern Hemisphere its path would appear as a curve to the *left* of the path it would have taken if the earth had not been rotating. This can be summarised by stating that in the N. Hemisphere winds and currents are deflected to the *right* whilst in the S. Hemisphere they are deflected to the *left*.

The figure (below, left) shows deflection to the right of an air or water mass moving (i) towards the pole, and (ii) towards the equator in the Northern Hemisphere.

The figure (below, right) shows deflection of an air or water mass moving (i) towards the pole, and (ii)towards the equator in the Southern Hemisphere.

Revolution of the Earth

The earth takes $365\frac{1}{4}$ days to revolve once round the sun. Every fourth year is given 366 days and this is called a *leap year*. All other years have 365 days. The earth's axis always points in the same direction in the sky, i.e. to the Pole Star. It is also permanently tilted at an angle of $66\frac{1}{2}°$ to the earth's Orbital Plane. The revolution of the earth and the inclination of its axis result in:

(i) Changes in the altitude of the mid-day sun at different times of the year

(ii) Varying lengths of day and night at different times of the year

(iii) The four seasons.

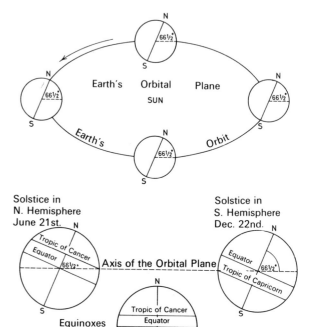

Changing Altitudes of the Mid-day Sun at Different Times of the Year

Summer solstice June 21

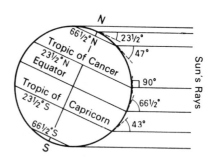

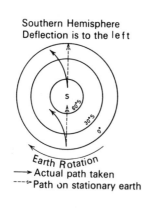

Northern Hemisphere
Deflection is to the right

Southern Hemisphere
Deflection is to the left

→ Actual path taken
---→ Path on stationary earth

Earth Rotation

10

Equinoxes Mar. 21, Sept. 23

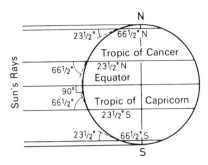

Winter solstice Dec. 22

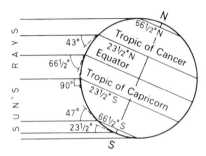

The apparent path of the sun between sunrise and sunset for selected latitudes during the solstices and the equinoxes

(I) North Pole

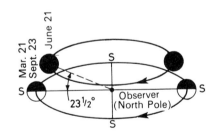

March 21 Sun circles the Pole, one half of it being visible above the horizon.

June 21 After March 21 sun rises higher in sky and is visible 24 hours each day. Highest altitude of sun is on June 21.

Sept. 23 Sun's path the same as for March 21. Sun is visible from March 21 to Sept. 23.

Dec. 22 After Sept. 23 sun is not visible above the horizon until March 21.

(II) Arctic Circle

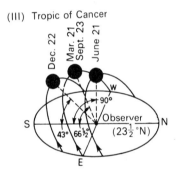

March 21 Sun rises due east and sets due west. It is visible for 12 hours.

June 21 Sun is visible for 24 hours.

Sept. 23 Sun rises and sets as for March 21.

Dec. 22 Sun is only visible for a few minutes when it appears above the southern horizon.

(III) Tropic of Cancer

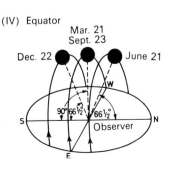

March 21 Sun rises due east and sets due west. It is visible for 12 hours.

June 21 Sun rises north of east and sets north of west. At noon its altitude is 90°.

Sept. 23 Sun rises and sets as for March 21.

Dec. 22 Sun rises south of east and sets south of west. It is visible for less than 12 hours.

(IV) Equator

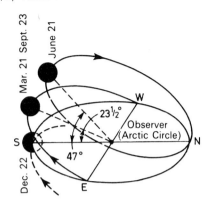

The range of mid-day altitudes of the sun for selected latitudes.

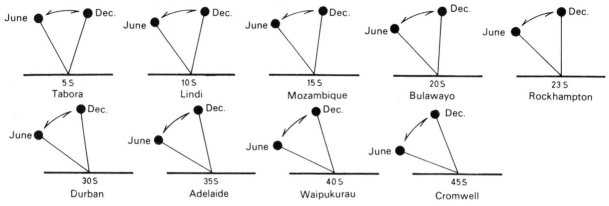

5 S	10 S	15 S	20 S	23 S
Tabora	Lindi	Mozambique	Bulawayo	Rockhampton

30 S	35 S	40 S	45 S
Durban	Adelaide	Waipukurau	Cromwell

March 21   Sun rises due east and sets due west. Mid-day altitude is 90°.

June 21   Sun rises north of east and sets north of west.

Sept. 23   Sun rises and sets as for March 21.

Dec. 22   Sun rises south of east and sets south of west.

Note   The sun is visible for 12 hours every day of the year.

The Varying Lengths of Day and Night at Different Times of the Year

The shaded part of each diagram represents night. The lengths of day and night for a selected parallel can be found by comparing that part of the parallel in the shaded zone with that part of it in the non-shaded zone.

In each diagram one half of the equator has night while the other half has day, i.e. DAY ≡ NIGHT along the equator throughout the year.

Throughout the year one half of the earth has day while the other half has night. Only during the equinoxes does the dividing line between day and night coincide with meridians (see diagram below). During the equinoxes the sun is overhead at noon along the equator and at these times (March 21 and September 23) DAY ≡ NIGHT along every parallel.

On June 21 the sun is overhead at noon along the Tropic of Cancer and all parallels in the Northern Hemisphere have their longest day of the year. At this time the length of the day increases as latitude increases north of the equator until there is continuous day north of the Arctic Circle. South of the equator the length of the day decreases with increasing latitude until there is continuous night south of the Antarctic Circle.

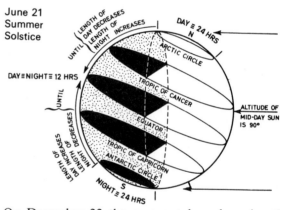

On December 22 the reverse takes place, i.e. the length of the day increases with increasing latitude south of the equator but decreases with increasing latitude north of the equator.

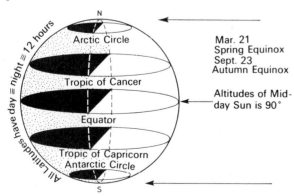

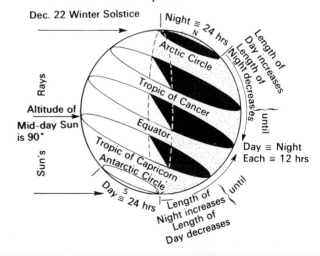

The Seasons

All part of the earth's surface except the equatorial latitudes experience a definite rise in temperature during one part of the year and a corresponding fall in temperature during another part of the year. This rise and fall in temperature is chiefly caused by the varying altitude of the mid-day sun and the number of hours of daylight. High mid-day sun altitudes cause high temperatures whereas low mid-day sun altitudes cause low temperatures.

The diagram below shows three bands of light (A, B, C) each containing the same amount of sun energy (this is indicated by equal diameters shown in dotted lines). Band A has its energy spread over Surface A; Band B has its energy spread over Surface B, and Band C has its energy spread over Surface C. Clearly the temperature will be highest at A and lowest at C.

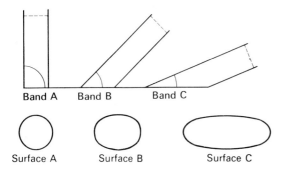

Sun's altitude in relation to heating

The monthly positions of the earth in its revolution round the sun are shown in the diagram below. The parallel at which the sun is overhead at noon is shown for each position. The overhead position of the sun 'moves' from the equator on March 21 northwards to the Tropic of Cancer on June 21, then back to the equator on September 23 and then 'moves' southwards to the Tropic of Capricorn on December 22 and finally returns to the equator on March 21.

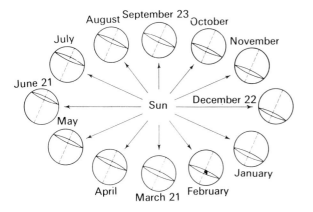

Associated with the overhead sun is a belt of heat and it is the movement of this belt between the Tropics which results in the alternation of the seasons. The Northern Hemisphere receives its maximum amount of solar radiation during June and its minimum amount during December (diagrams on page 12 and below). Between March 21 and September 23 this hemisphere has its summer while between September 23 and March 21 it has its winter. Spring and Autumn are two shorter seasons which occur between the two main seasons and which represent a transition from one to the other.

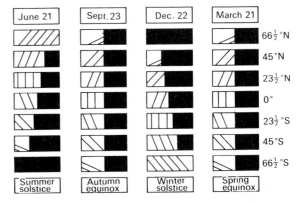

Each rectangle represents 24 hours. Night is shaded and day is left white.

The lines represent the altitude of the sun at mid-day.

The relationship between length of day, latitude and time of year is shown in the diagram above. Also shown is the altitude of the mid-day sun for each latitude. Notice how the length of day decreases from a maximum at $66\frac{1}{2}°$N to a minimum at $66\frac{1}{2}°$S on June 21, and how it increases from a minimum at $66\frac{1}{2}°$N to a maximum at $66\frac{1}{2}°$S on December 22. Finally notice how all latitudes have day equal to night at the equinoxes and how day always equals night along the equator.

EXERCISES

Draw a diagram to illustrate the revolution of the earth around the sun by showing the positions of the earth at the two equinoxes and the two solstices. On your diagram insert the following:
 (i) the path in orbit
 (ii) an arrow showing the direction of the earth's movement along the orbit
 (iii) an arrow to show the direction of the earth's rotation at one position of the earth
 (iv) the equator and the two tropics
 (v) the date at each position.

Objective Exercises

1 The distance of the Earth from the Sun is about
  A 1500 million km
  B 300 million km
  C 227 million km
  D 149 million km
  E 120 million km

  A B C D E
  ☐ ☐ ☐ ☐ ☐

2 What is the name of the planet which takes 88 days to make one revolution of the Sun?
  A Venus
  B Mercury
  C Earth
  D Pluto
  E Neptune

  A B C D E
  ☐ ☐ ☐ ☐ ☐

3 Which one of the following is **not** connected with proofs of the Earth's shape?
  A rotation and revolution
  B circumnavigation
  C the Earth's shadow on the moon during an eclipse
  D the Bedford Level Canal Experiment

  A B C D
  ☐ ☐ ☐ ☐

4 What is meant by the eclipse of the Moon?
  A It is the path along which the moon revolves.
  B When the Moon comes between the Sun and the Earth it causes the shadows of the Moon to fall on the Earth.
  C It occurs when the Earth comes between the Sun and the Moon and the centres of all three are on the same straight line.
  D For any place, it is the average angle made by a line drawn from the Moon to a place and the horizon at midnight.

  A B C D
  ☐ ☐ ☐ ☐

5 Which of the following statements can be taken as evidence to show that the Earth is spherical?
  A The rotation of the Earth from west to east.
  B Some parts of the Earth have day when other parts have night.
  C The Earth's horizon is seen to be curved when seen from an aeroplane.
  D The Earth's revolution around the Sun.

  A B C D
  ☐ ☐ ☐ ☐

6 The Earth makes one complete revolution of the Sun in
  A 365 days
  B 360 days
  C 365 $\frac{1}{4}$ days
  D 1 day

  A B C D
  ☐ ☐ ☐ ☐

7 An eclipse of the Sun, takes place
  A when the Moon passes between the Sun and the Earth
  B once every five years
  C when the Moon is full
  D when the Earth comes between the Sun and the Moon

  A B C D
  ☐ ☐ ☐ ☐

8 This question is based on the diagram given below. The position of the overhead Sun at noon indicates that this diagram shows the Earth's orbital position on
  A 22nd December
  B 21st March
  C 21st June
  D 23rd September

  A B C D
  ☐ ☐ ☐ ☐

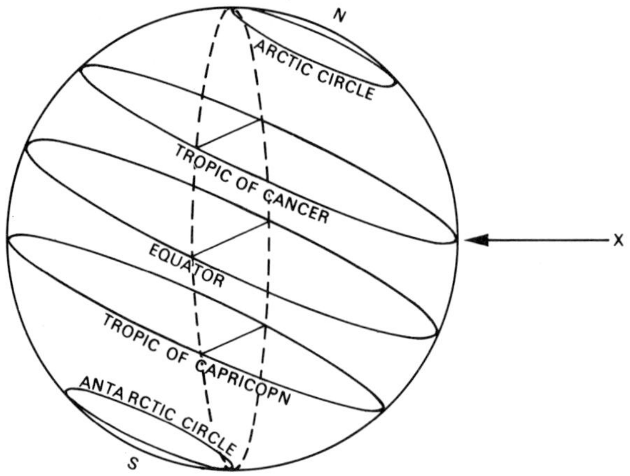

9 Which of the following statements best describes longitude?
  A An imaginary line on the Earth's surface joining the North and South Poles.
  B The angular distance east or west of the Greenwich Meridian.
  C The distance of a place east or west of the Greenwich Meridian.
  D The position of a place on the Earth's surface with reference to the Prime Meridian.
  E A line on a map that cuts the Equator at right angles.

  A B C D E
  ☐ ☐ ☐ ☐ ☐

10 The permanent tilt of the Earth's axis and the revolution of the Earth in its orbit together cause

A day and night

B varying lengths of day and night at different times of the year

C differences in time between places on different meridians

D the deflection of winds

A B C D
□ □ □ □

11 At the summer solstice in the Northern Hemisphere, which of the following latitudes will have the longest night?

A 45°N

B $23\frac{1}{2}$°S

C 66°N

D 66°S

A B C D
□ □ □ □

12 When the Sun is vertically overhead along the Tropic of Capricorn at midday

A days and nights are of equal length in the Northern Hemisphere

B nights are longer than days in the Southern Hemisphere

C days and nights are of equal length at the Poles

D night is equal to 24 hours at the North Pole

A B C D
□ □ □ □

13 Which one of the following is **not** connected with proofs of the Earth's shape?

A rotation and revolution

B circumnavigation

C the Earth's shadow on the moon during an eclipse

D the Bedford Level Canal Experiment

A B C D
□ □ □ □

14 What is meant by the eclipse of the Moon?

A It is the path along which the moon revolves.

B When the Moon comes between the Sun and the Earth it causes the shadows of the Moon to fall on the Earth.

C It occurs when the Earth comes between the Sun and the Moon and the centres of all three are on the same straight line.

D For any place, it is the average angle made by a line drawn from the Moon to a place and the horizon at midnight.

A B C D
□ □ □ □

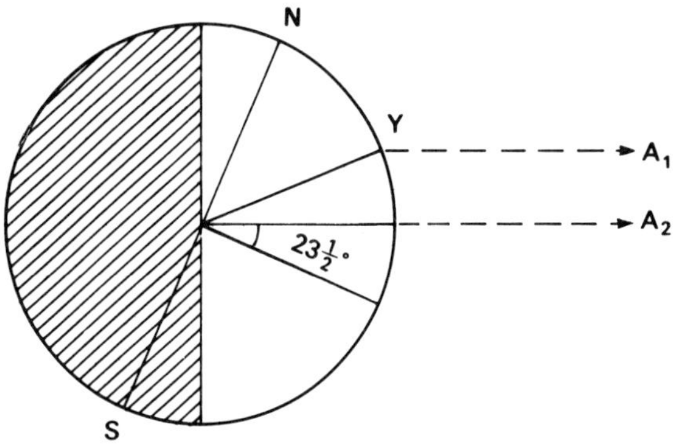

15 The above diagram shows the Sun's rays A_1 and A_2 shining on the Earth on 21st June. The latitude of Y, a point on the Earth's surface, is 45°N. What is the elevation of the Sun at Y?

A $23\frac{1}{2}$°

B $21\frac{1}{2}$°

C $68\frac{1}{2}$°

D 45°

A B C D
□ □ □ □

16 Which of the following statements can be taken as evidence to show that the Earth is spherical?

A The rotation of the Earth from west to east.

B Some parts of the Earth have day when other parts have night.

C The Earth's horizon is seen to be curved when seen from an aeroplane.

D The Earth's revolution around the Sun.

A B C D
□ □ □ □

15

2   The Earth's Crust

STRUCTURE OF THE EARTH

Although man has not been able to sink boreholes more than a few kilometres into the earth's crust, he is able to obtain information on the nature of the earth's interior by studying lavas emitted from volcanoes and by studying the behaviour of earthquake waves. From these studies it appears that the earth is composed of three parts, the core, the mantle and the crust as shown in this diagram. Our knowledge of the earth's interior increases year by year and this knowledge is used to modify our theories about the earth's interior. Until recently it was thought that the earth's outer crustal layer was composed of rafts of SIAL (light rocks rich in silica and alumina) which 'floated' on a 'sea' of SIMA (heavier rocks rich in silica and magnesia). But recent studies have revealed that large areas of the outer crustal layer are made of basaltic rocks similar to the SIMA. For example, there are vast areas of newly formed basaltic rocks which form the ocean bed in the central parts of the Atlantic and Indian Oceans. Similar rocks form the ocean bed of the waters around Antarctica. It is thought that these basaltic rocks emerge from the earth's interior near to the mid-oceanic ridges and slowly push outwards away from these ridges. The earth's crust is now regarded as a series of *plates* which are gradually being pushed apart, away from the zone where they formed. The sial rocks carried on some of the plates form the continents.

The plates eventually collide and when this happens one of them is drawn underneath another to make room for the newly forming plates. There is evidence which suggests that this is happening in *zones of subduction* along the edges of the continental plates. Active zones of subduction occur off the coasts of Japan, California and South America, and in the Caribbean area and around New Zealand. The map on page 17 shows the locations of the earth's main plates in relation to the oceanic ridges.

The earth, which is thought to have been formed about 6000 million years ago, has experienced a great many changes. Although these changes take place very slowly, there is evidence that considerable changes in the shape and character of the land masses have taken place since man first appeared.

Distribution of Land and Water

The water surface of the earth accounts for just over 70% of the earth's surface. This distribution is not the same for both Northern and Southern Hemispheres. In the latter it is as much as 80% of the total surface area of that hemisphere. The

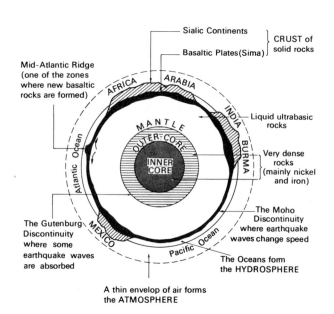

Section through the Earth at 20 °N

average height of the land is just under 900 metres (2950 feet) while the average depth of the oceans is 3800 metres (12 400 feet). See the diagram below. The greatest height is Mt. Everest (8875 metres or 29120 feet) on the northern border of Nepal, and the greatest depth is the Mindanao Deep (10 490 metres or 34 400 feet) off the east coast of the Philippines.

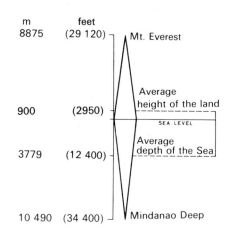

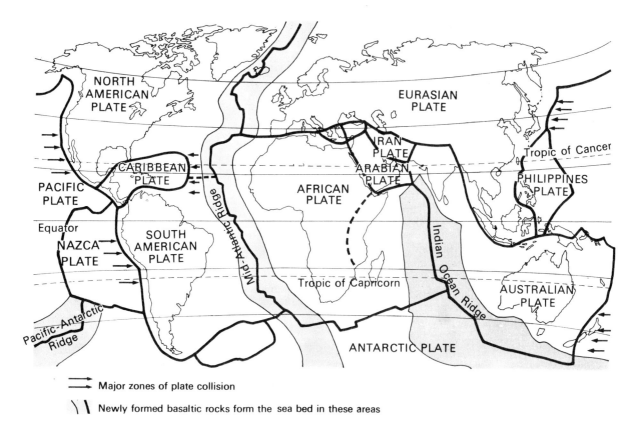

The Earth's Crust is composed of Plates

Legend:

⟶ Major zones of plate collision

\ \ Newly formed basaltic rocks form the sea bed in these areas

The relationship between land and water surfaces is shown by the diagram on the right. Two main levels can be recognised: (i) the *ocean floor* level, and (ii) the *continental level*. The two are connected by the *continental slope*. The edge of the continental level is submerged to a depth of about 200 metres (650 feet) and this zone is called the *continental shelf*. The seas on this shelf are called *epicontinental* or *shelf seas* (the importance of these is discussed in a later section). The more important of these shelf seas in the Tropics are (i) between N. Australia and New Guinea, (ii) between Borneo and the Malay Peninsula and Thailand, and (iii) along the Gulf Coast of North America.

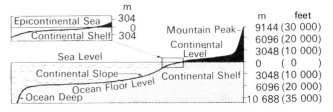

The continental surface is broken by mountain ranges, plateaus, and plains, whilst the ocean floors are far from level. Basins and deeps cause the floor to plunge to great depths. Extending across most of the oceans are ridges, some of which rise above the level of the sea to form chains of islands. Extensive plateaus also occur in some oceans.

ROCKS AND MINERALS

The earth's crust is composed of rocks each of which is made up of minerals. Most minerals are compounds of several elements, e.g. silica (SiO). A few minerals are themselves elements, e.g. carbon (diamond), gold and sulphur. Silica often combines with other oxides to form *silicates*, the most common of which are *felspars*. *Mica* is another common silicate.

Minerals are frequently crystalline, i.e. the atoms forming the crystals are arranged in a definite manner. Some minerals are non-crystalline, i.e. the atoms forming the mineral are not arranged in any definite order.

Felspars are silicates of aluminium (Al), potassium (K), sodium (Na) and calcium (Ca).

Augite, *hornblende* and *olivine* are silicates of iron (Fe), magnesium (Mg), calcium (Ca) and aluminium (Al).

Clay minerals are complex silicates derived from

17

weathered minerals such as felspars. They are silicates of aluminium (Al).

Felspars which weather under tropical humid conditions lose their silica. The residue is chiefly oxides of aluminium and these are called *bauxite*.

Classification of Rocks according to origin

IGNEOUS — These rocks have been formed inside the earth, under great pressure and heat. They do not occur in layers and most of them are crystalline. Some rocks, e.g. granite, have cooled slowly and contain large crystals: others, e.g. basalt, have cooled quickly and contain small crystals. Some rocks contain a high percentage of silica and are called *acid rocks*. Granite is a good example. Other rocks, such as basalt, contain a high percentage of iron, or aluminium or magnesium oxides and are called *basic rocks*. Igneous rocks do not contain fossils.

SEDIMENTARY — The most common sedimentary rocks are composed of particles of rocks which have been deposited, usually in layers, by water, wind or moving ice. These are called *mechanically-formed*, or *elastic sedimentary rocks* which are largely or completely made of particles of organic matter, or they are made of minerals which have been chemically deposited. All sedimentary rocks are non-crystalline. They contain fossils.

METAMORPHIC — These are rocks whose structure and appearance have been changed by great heat, or great pressure, or both. Any rock can be changed into a metamorphic rock.

Igneous Rocks

There are two main groups.

I *Volcanic* (these have been poured out onto the earth's surface, and they are called *lavas*) e.g. basalt

II *Plutonic* (these have solidified deep in the earth's crust and they reach the surface only by being exposed by erosion) e.g. granite

Sedimentary Rocks

There are three main groups.

  I *Mechanically-formed*
    (i) Wind-deposited e.g. loess
    (ii) River-deposited e.g. clays, gravels and alluviums
    (iii) Glacier-deposited e.g. moraines, sands and gravels and boulder clay
    (iv) Sea-deposited (very similar to those of (ii)

  II *Organically-formed*
    (i) From animals e.g. chalk and coral
    (ii) From plants e.g. peat, lignite, coal

 III *Chemically-formed*
    e.g. rock salt, borax, gypsum, nitrates, potash and certain limestones

Metamorphic Rocks

e.g. marble (from limestone)
    slate (from clay)
    gneiss (from granite)
    quartzite (from sand)
    graphite (from coal)

Grouping of sedimentary rocks according to their texture

CLAYS — Composed of microscopically fine particles

SILTS — Composed of particles not quite so fine

SANDS — Composed of coarser particles (easily seen with the naked eye). When cemented together they form *sandstones* (these are called *grits* when the sand grains are angular)

GRAVELS — Composed of rounded and large particles. When cemented together they form *conglomerates*

BRECCIA — Composed of coarse and angular particles which have been cemented together

JOINTS

Rocks very often develop cracks when they are subjected to strain produced by compression or tension. The strain may be caused by earth movements, or by contraction when molten rocks solidify, or by the shrinking of sedimentary rocks on drying. The cracks so formed are called *joints*. In sedimentary rocks joints are often at right angles to the bedding plane. Sometimes more than one set of joints develops. When this happens the rock becomes broken into blocks, e.g. limestone and sandstone, or into columns as in some lavas, e.g. basalt.

Sandstone cliff showing bedding planes

Note Jointing does not result in the displacement of rocks as does faulting (see pages 21 and 22).

Horizontal bedding planes in cliffs on the coast of South-West England

Joints in granite rocks in South-West England

THE FORCES WHICH PRODUCE PHYSICAL FEATURES

A. Internal (operate within the earth's crust)

I Earth Movements

(i) *Vertical* — (up and down) movements cause faulting of the crustal rocks. Features produced: *plateaus*; *block mountains* (horsts); *basins*; *some types of escarpment.*

(ii) *Lateral* — (sideways) movements cause folding of the crustal rocks. Features produced: *fold mountains*; *rift valleys*; *horsts, block mountains*

II Volcanic Eruptions

(i) *External* — (lavas reach the earth's surface). Features produced: *lava plains and plateaus*; *volcanic cones*; *geysers*

(ii) *Internal* — (lavas solidify in the crust). Features produced: *dykes*; *sills*; *batholiths*; *laccoliths*

B. External (operate on the earth's surface)

I Denudation

(i) *Weathering* — (the break-up of rocks by alternate heating and cooling; chemical actions; and the action of living organisms). Features produced: *soil*; *earth pillars* (by rain action); *screes*

(ii) *Erosion* — (the break-up of rocks by the action of rock particles being moved over the earth's surface by water, wind and ice). Features produced: *valleys*; *peneplains*; *cliffs*; *river* and *coastal terraces*; *escarpments*

(iii) *Transport* — (the movement of rock particles over the earth's surface by water, wind and ice)

II Deposition

(i) *By Water* — (river or sea). Features produced: *flood plains*; *levées*; *alluvial fans*; *deltas*; *beaches*; *lake plains*; *marine alluvial plains*

(ii) *By Ice* — (ice sheets and valley glaciers). Features produced: *boulder clay plains*; *outwash plains*; *moraines*; *drumlins*; *eskers*

(iii) *By Wind* — Features produced: *loess plains*; *sand dunes*

(iv) *By Living Organisms* (e.g. coral) Features produced: *coral formations*

(v) *By Evaporation and Precipitation* Features pro-

(vi) *Of Organic Matter*

duced: *salt deposits*

Features produced: *coal deposits*

Internal Forces
Earth movements

The face of the earth is full of features, some small and some large, all of which are being slowly but constantly changed by the agents of denudation – rivers, waves, wind, moving ice, frost, rain and sun. The major features such as mountains, plateaus and plains have been formed by earth movements. These movements, which are lateral and vertical, exert great forces of tension and compression, and although they usually take place very slowly, they eventually produce very impressive features. Before studying the effects of earth movements it is as well to get to know the meanings of some common terms. Sedimentary rocks are formed from sediments which have been laid down in horizontal layers. This layering is called *stratification*, and sedimentary rocks are therefore *stratified rocks*. The face of each layer is called the *bedding plane*.

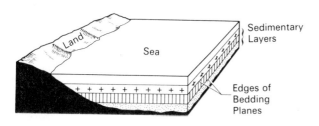

Earth movements cause sedimentary rocks to be displaced, i.e. to be pushed out of the horizontal plane so that the rocks are *tilted* or *inclined*. The inclination of the rocks is called the *dip*. The direction parallel to the bedding plane and at right angles to the dip is called the *strike*.

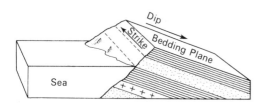

Earth movements can also cause *folding* and *faulting* of the sedimentary rocks. Folding results from *lateral forces of compression*.

The Nature of Folds

The diagrams show a *simple fold*. The layers of rock which bend up form an *upfold* or *anticline*. Those which bend down form a *downfold* or *syncline*. The sides of a fold are called the *limbs*.

If compression continues then simple folds are changed first into *asymmetrical folds*, then into *overfolds*, and finally into *overthrust folds*.

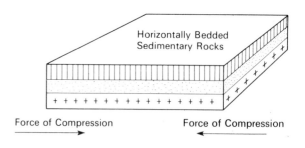

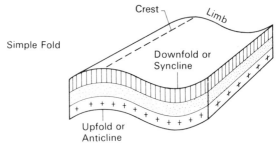

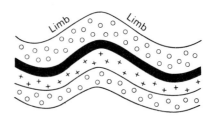

Simple Fold

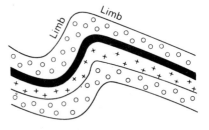

Asymmetrical Fold

Asymmetrical fold: one limb steeper than the other

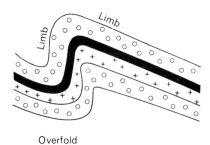

Overfold

Overfold: one limb is pushed over the other limb

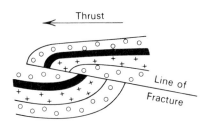

Overthrust Fold

Overthrust fold (nappe) when pressure is very great a fracture occurs in the fold and one limb is pushed forward over the other limb.

Folded rocks in South-West England

The Nature of Faults

Faulting can be caused by either *lateral* or *vertical* forces of either *compression* or *tension*. Tension causes a *normal fault* (right, top) and compression causes a *reverse fault* (right, bottom).

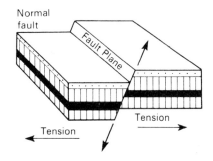

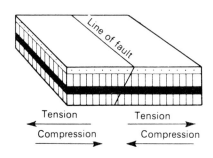

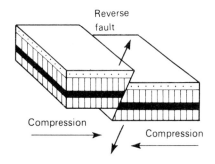

Forces of Tension and Compression

Rocks of the earth's crust are subjected to tension and compression when vertical or lateral earth movements take place. If one part of the crust is compressed then clearly another part must be stretched, i.e. be under tension. Rocks when under tension usually fault, but when under compression they may fold or fault depending on whether they are brittle or flexible under stress.

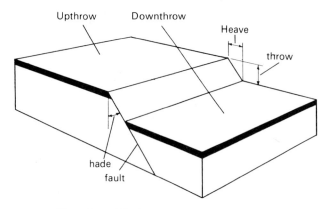

The nature of Faults

Upthrow	upward displacement
Downthrow	downward displacement
Throw	vertical displacement
Heave	lateral displacement
Hade	inclination of the fault to the vertical

Normal Fault (caused by tension)

(i)

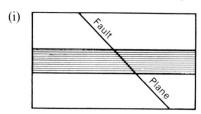

Fault plane develops

(ii)

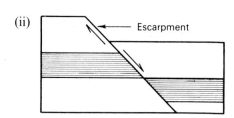

Rocks on each side of the fault plane are usually displaced as shown by the arrows.

(iii)

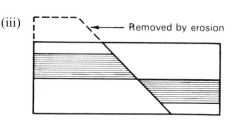

Faulting sometimes produces an escarpment. This is sometimes removed by erosion.

Note Surface area is increased.

Reverse Fault (caused by compression)

(i)

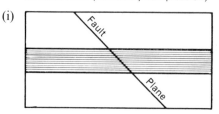

Fault plane develops.

(ii)

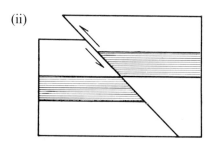

Rocks are displaced as shown by the arrows. Those on one side of the fault plane ride up over those on the other side.

(iii)

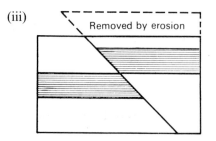

An escarpment may mark the fault. Erosion may later remove this.

Note Surface area is reduced.

Tear Fault

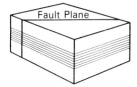
Fault Plane

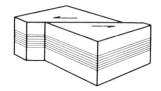

The fault plane is often almost vertical. The rocks are displaced horizontally as shown by the arrows. There is no vertical displacement.	Tear faults usually occur during earthquakes. The earthquakes which wrecked San Francisco produced tear faults e.g. the San Andreas Fault.	Note: Earthquakes sometimes produce vertical movements or a combination of vertical and horizontal movements e.g. Midori, Japan.

Vertical Movements

There is evidence in many countries to show that these movements take place. There are beaches in Scotland and Norway which now lie several feet above the sea and there are submerged forests around the coasts of Britain and other countries.

Horizontal Movements

There is considerable evidence which suggests that the surface of the earth has been moving, and that it is still moving horizontally. We have seen that new rocks appear to be forming around the mid-oceanic ridges and as these move outwards, older rocks are drawn down into the mantle in the zones of subduction, which is where crustal plates collide. Wegener was the first person to suggest that the positions of the continents had changed, and he suggested that originally there was only one continent, which he called PANGEA. He postulated that this was located around the South Pole during the Carboniferous Period (see the Table on the inside back cover), and that during the early Tertiary Period, this ancient continent started to split up into several parts. Nobody has yet explained why this should have happened, or what forces could have been powerful enough to set such an enormous change in motion. It may have been caused by strong convection currents forming in the mantle, but no reason can be found to explain why such currents, powerful enough to start what is known as *Continental Drift,* should have started. The Himalayas, the Rockies and the Andes were formed at about this time, and Wegener believed that they originated through 'crumpling' at the edges of the continents as they drifted away from the South Pole, northwards, eastwards and westwards.

There is a considerable variety of evidence in all the continents which suggests that at one time the continents may have all been joined together. This evidence includes types of fossils, both plant and animal, and rock types and rock structures. The shapes of the continents suggest that they can be fitted together. This is especially true of the Americas, Europe and Africa. A group of scientists recently studied the rocks in the Gulf of Guinea in West Africa, and by using a computer map they decided to visit a specific area on the coast of Brazil to see whether the rocks there had any similarity with those of Guinea. On visiting that part of the Brazilian coast they discovered that the rocks were identical.

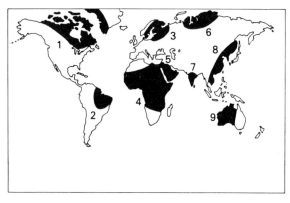

1	Laurentian Shield	6	Siberian Shield
2	Brazilian Shield	7	Deccan Shield
3	Baltic Shield	8	China Shield
4	African Shield	9	Australian Shield
5	Arabian Shield		

Pangea in the late Palaeozoic Period

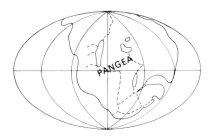

The early Tertiary Period (the continents begin to drift apart)

Note Europe and N. America were partly joined, forming *Laurasia*, while Africa was the ancient continent of *Gondwanaland*

The last Ice Age

The Continents Today (position of similar rocks across the Atlantic Ocean is shown)

＿ ＿ Main mid-ocean ridges where new rocks are being formed

INTERNAL AND EXTERNAL FORCES OPERATE TOGETHER

The theory of *isostacy* has been accepted for a long time to explain why vertical and horizontal movements take place in the earth's crust. This theory compares the continental blocks with corks floating on a liquid. If a thin layer is cut from the top of the corks then they will rise and the corks will not project as far down into the liquid as before. In a similar way the theory states that the light sial rocks of the continents are 'floating' on the denser sima rocks which are in a semi-liquid state. As the continents are worn down by denudation they gradually rise up. This concept may be applied to the theory of Plate Tectonics, which attempts to explain the destruction and renewal of crustal rocks, as follows. The zone of subduction is said to be a low-angle fault at the junction of two colliding plates. To relieve the pressure between the plates, the advancing plate slides down below the stronger plate, as shown in this diagram. As the rocks in the subsiding plate melt and mix with the magma at the base of the crust, the pressure frequently causes earthquakes, and volcanic mountains and islands often develop 150 to 400 kilometres (about 250 to 650 miles) away from the inclined fault. The lava emitted from these volcanoes may be a mixture of the old rocks, which were drawn down into the zone of subduction, and completely new rocks (magma) from the earth's interior. Many of the world's volcanic zones, for example, those of Japan, the Philippines, the Caribbean Islands, the Andes and the Rockies are near to zones of subduction. All of these volcanic zones experience earthquakes.

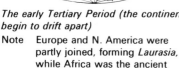

Before Erosion

Highland
Ocean Continent Ocean
MANTLE

▥ Sialic rocks form the continent

◼ Sima plate is thicker below the continent

Process of Erosion

Erosion
Deposition Deposition
MANTLE flow of mantle

◼▮◼ Pressure as a result of erosion and deposition

Balance is regained (isostacy)

Continent
Ocean Ocean
MANTLE

Note Sialic rocks are added to by volcanic activity

Line of volcanoes (or island arcs)
Ocean Deeps (Trenches) Zone of Subduction

CONTINENT

SEA
Basaltic Plate
Mantle

ADVANCING PLATE

Sialic and Sedimentary Rocks

Basaltic Plate carrying the Continent

Old crustal rocks are consumed and mixed with mantle to form **new magma**

Block diagram showing the Zone of Subduction and related Volcanic Mountains

Ice Sheets can cause the earth's crust to sag

1 The northern parts of North America and Europe lay under extensive ice sheets of vast thickness during the *Ice Age*. The weight of the ice caused the crust to be depressed.

2 Since the melting of these ice sheets, the crust has been slowly rising. The Scandinavian coasts have many sea beaches which lie from 8 metres to 30 metres (26 feet to 100 feet) above the present-day sea beaches. Evidence suggests that these *old sea beaches* have been raised because the land has been uplifted.

FEATURES PRODUCED BY EARTH MOVEMENTS

1 Fold Mountains

Fold mountains consist of great masses of folded sedimentary rocks whose thickness is often as much as 12 000 metres (40 000 feet) or more. Originally the sedimentary rocks must have been laid down in horizontal layers which later became folded through compression. To begin with, earth movements must have caused a part of the earth's crust to warp downwards into large depressions, called *geosynclines* which became the sites of seas and in which the sediments collected. In time the sediments caused further subsidence by their own weight. Also, the process of folding caused the width of the sedimentary rock zone to decrease and its thickness to increase further.

At one time many people thought that fold mountains developed through the contraction of the earth's crust on cooling but this explanation is not accepted today. The origin of fold mountains and other large landforms is much more complex than this. The explanation given today, though largely dependent on Holmes' theory, takes into account the theory of Plate Tectonics. The following diagrams suggest how fold mountains may have been formed. During the later stages of their formation, the fold mountains were faulted and igneous intrusions and volcanic features were added to the pattern of fold sedimentary rocks. Magma intrusions into the base of fold mountains formed batholiths (see Chapter 3), and some of these now lie exposed at the surface as a result of prolonged denudation which has removed the overlying rocks.

Types of Fold Mountains

Sometimes the folding is never intense and this simple folding gives rise to mountains and valleys. The anticlines become the mountains and the synclines the valleys as in the Jura of France. Simple fold mountains are rare.

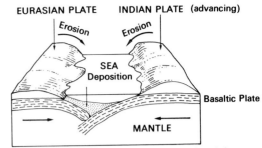

EURASIAN PLATE   INDIAN PLATE (advancing)

Colliding plates cause depression of the land and the formation of a shallow sea. Layers of sediments accumulate on the bed of this sea.

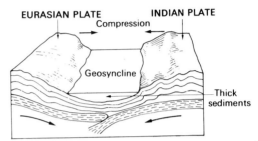

Sedimentary layers occupy a large depression or geosyncline and they become folded as the plates move together

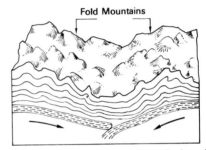

The sea has drained away and the rocks have been more intensely folded

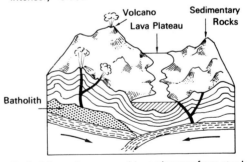

Batholiths, volcanoes and lava plateaus form at a later stage

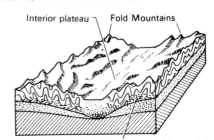

Volcanic rocks form the base of the fold mountains.

Complex folding is much more common. When this happens there is little or no relationship between the anticlines and the mountains, and the synclines and the valleys.

The peaks and valleys which occur in complex fold mountains have been formed by glacial and, or, river erosion (this will be dealt with later).

Some complex fold mountains do not appear to have been formed between two continents, e.g. the mountains of Java and Sumatra. The floor of the ocean probably acted as the 'missing' continent.

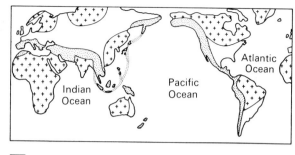

| Young Fold Mountains | Ancient Shields |

World Distribution of Young Fold Mountains

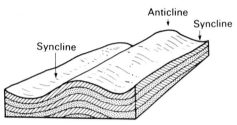

1 *Anticline and synclines*

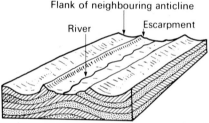

2 *Anticline opened by erosion (neighbouring anticlines will be opened up later by same process)*

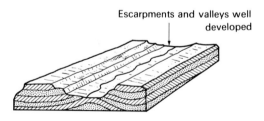

3 *Anticlinal valley develops rapidly as river excavates weak strata*

4 *Synclines stand up as mountains*

The Fate of Fold Mountains

Weathering and erosion attack fold mountains as they begin to form. The building of mountains like the building of all other major features of relief takes vast periods of time to complete. From the beginning of their formation weathering and erosion, operating together, attack and wear away the mountains. To begin with, earth movements are more powerful than weathering and erosion and the mountains reach heights of thousands of metres, but as earth movements weaken the wearing away process becomes dominant. In time they are reduced to an almost level surface not far above sea level. This surface is called a *peneplain*.

In this process of mountain reduction, strange things can happen. Rocks which were contorted into synclines are sometimes turned into mountains by river erosion. This is achieved by rivers opening up the weakened rocks of neighbouring anticlines which in time are completely removed leaving the adjoining synclines as mountains. The diagrams on the left explain this.

How many times have Fold Mountains formed on the earth's surface?

There is evidence to show that there have been three main mountain-building periods during the last 400 million years. The remnants of the first of these can be seen in the Altai Mountains, the Scandinavian Mountains and the Lake District of England. Remnants of the second mountain-building period occur in the Tien Shan Mountains and in the plateaus of central France and Spain. As far as present-day mountain systems go it is the last mountain-building period which is the most important. Mountains belonging to this period include the Himalayas, Andes, Rockies and Atlas Mountains. They are all called *Young Fold Mountains* (see diagram at top of page).

Is Mountain building still going on?

We know that as fold mountains are worn away the deposition of sediments in neighbouring oceans gives rise to new sedimentary rocks. Scientific surveys show that there is an extensive geosyncline forming from south of the Atlas Mountains to south of Indonesia. The geosyncline is represented by the Shotts of Tunisia and Algeria, the Mesopotamian Alluvial Plain, the Persian Gulf, the Indo-Gangetic Plain and the ocean deep south of Java. The stage of mountain building reached in the geosyncline varies from one part of it to another. In the eastern part there is frequent volcanic activity.

Influence of Fold Mountains on Human Activities

1 They often act as climatic barriers. Regions on one side of a mountain range may have an entirely different climate from that of the region on the other side. The coastlands of British Columbia have mild winters, warm summers and rain throughout the year. To the east of the Rockies the prairies have cold winters, hot summers and there is a maximum of rain in the summer.
2 They often receive heavy rain and/or snow which may give rise to important rivers. Most of the rivers of Asia rise in the mountain ramparts of central Asia. These rivers may be used for irrigation, e.g. the Ganges and Indus, or they may be used for developing hydro-electric power (H.E.P.), e.g. the Colorado and Columbia and the rivers of Switzerland and Japan, or they may be used for both irrigation and developing H.E.P., e.g. Murray River in S.E. Australia.
3 Some mountains and their plateaus may contain minerals, e.g. Nevada (copper and gold), Bolivia (tin).
4 They may act as barriers to communications or they may make the construction of communications difficult.
Some mountain ranges have valuable timber resources, e.g. coast ranges of Western America (coniferous soft woods), foothills of the Himalayas (teak).

2 Basins and plateaus

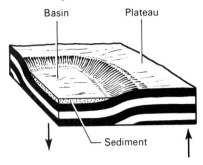

Vertical earth movements can cause the crust to warp, and sometimes large areas of it are uplifted whilst others are depressed. The uplifted areas form *plateaus*, sometimes called *tectonic plateaus*, and the depressed areas *basins*. There are two types of tectonic plateaus. Some plateaus slope down to surrounding lower land. The Deccan Plateau of India is an example. Other plateaus slope up to surrounding mountains. These are called *intermont plateaus* and the Tibetan and Bolivian Plateaus are examples.

Basins, too, are of different types. Some are *sea-basins*, e.g. the Celebes Sea (Indonesia) and the Black Sea (U.S.S.R.); others are high above sea level and rimmed by mountains, e.g. Lake Victoria Basin (East Africa), the Great Basin of Nevada (U.S.A.) and the Tarim Basin (central Asia); and others are filled with sediments and either have an external drainage, as in the Zaire Basin (central Africa), or are *basins of inland drainage*, as in the Chad Basin (north-central Africa) and the Tarim Basin (central Asia).

On page 28 is a diagrammatic section across the Western Highlands of the U.S.A. The Colorado Plateau is an intermontane plateau and lies between the Rockies and the Wasatch Mountains. The Great Basin is really a plateau which has been block-faulted (see second diagram). Block sections of the Plateau now form block mountains. Some of the depressions of the Great Basin have no

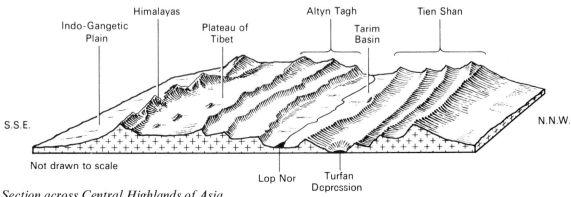

Section across Central Highlands of Asia

27

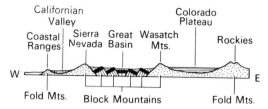

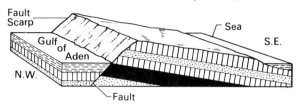

Tilt-Block Mountain of Somalia (E. Africa)

external drainage and because of a high rate of evaporation salt lakes frequently form. Some of these dry up and form salt flats or *playas*.

Central Asia contains numerous plateaus and basins. Tibet is the highest plateau in the world and is an intermontane plateau. The Turfan Depression lies below sea level while the Tarim Basin is between 600 and 2000 metres (1950 and 6550 feet) above sea level.

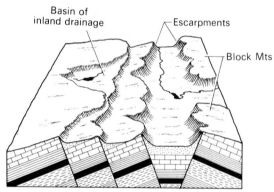

Block Faulting in the Great Basin (U.S.A.)

Horsts, Tilt Blocks and Rift Valleys

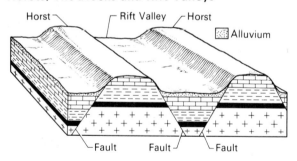

Earth movements sometimes cause the crust to be divided into rectangular-shaped blocks some of which are uplifted and others are depressed. This is called *block-faulting*. Uplifted blocks may either be tilted when they form *tilt blocks* or they may be horizontal to form *horsts*. The tilt block usually has one pronounced scarp, e.g. the Western Ghats of the Deccan, while horsts have two pronounced scarps.

Tilt Blocks	*Horsts*
Deccan Plateau	Korea
Arabian Plateau	Sinai
Brazilian Plateau	Black Forest

When blocks of the crust are depressed between parallel faults they often form *rift valleys*.

East African Rift Valleys

The most outstanding belt of rift valleys extends for 4800 kilometres (3000 miles) from Syria to the River Zambesi in East Africa. The belt contains a number of well-developed rift valleys some of which contain lakes whose beds are below sea level. Many of the valleys have precipitous sides which are fault scarps and which are often very straight for many miles. The broken line on the map indicates the main belt of rift valleys.

The River Rhine in Europe flows through an impressive rift valley between the fault blocks of the Black Forest and the Vosges.

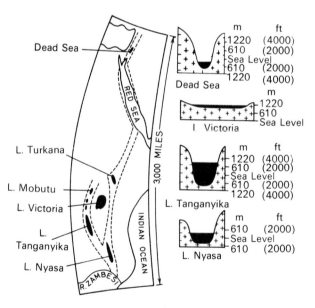

Note The level of the Dead Sea is 393 m (1290 ft) below sea level, and its floor is 819 m (2680 ft) below sea level.

Origin of Rift Valleys

They are thought to have developed either from the action of tensional forces in the crust which caused fault blocks to sink between parallel faults, or from the action of compressional forces in the crust which caused fault blocks to rise up towards each other and over a central block. Many people think that compression has been responsible for most rift valleys. They argue that it would not be possible for blocks of the crust to sink into the heavier rocks of the sima below the crust, unless they were trapped there by later pressure.

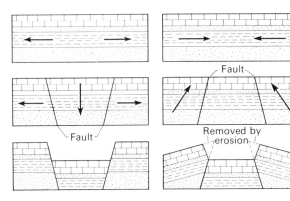

Formed by Tension
Layers of rocks are subject to tension.
Faults develop and the centre block begins to subside.

Formed by Compression
Layers of rocks are subject to compression.
Faults develop and the outer blocks begin to thrust up over the centre block.
The over-hanging sides of the rift valley are worn back by erosion.

After subsidence a depression with steep fault scarp sides, i.e. a rift valley, is formed. It is trapped in position by later pressure.

Rift Valley Formation

Recent research also suggests that some rift valleys, e.g. the Red Sea Rift and the rift in the Mid-Atlantic Ridge, were formed from one fault only, and not from two parallel faults. Yet their structure, with a depressed central portion and high horsts on either side cause them to be classified as rift valleys. The horsts of these rifts are maintained at a high level by the divergence of two plates at the rift and the subsequent replacement of the diverging rocks by newly formed basaltic rocks from the rift. It can be seen that rift valleys may form in different ways, and even within one rift valley the structure may vary in different parts of its length. The East African Rift has definite reverse faults in some places, while in other places the faults appear to be normal. This rift extends north to the Red Sea, which is thought to have been formed from a single fault in a zone of divergence. In addition to the very large rift valleys discussed here, there are many other, though smaller, rift valleys.

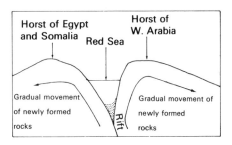

Influence of Plateaus on Human Activities

1 Some plateaus are rich in mineral deposits, e.g. the Highlands of Brazil (iron ore and manganese); the African Plateaus (copper and gold); Western Australia (gold).

2 High plateaus in tropical latitudes are sometimes of agricultural value, e.g. the Kenya Plateau (coffee, sisal and maize); the Brazilian Plateau (coffee).

Jordan Rift Valley

EXERCISES

1 With the aid of well-labelled diagrams, explain the differences in origin and appearance of the following:
 (i) faults and folds
 (ii) block mountain and rift valley
 (iii) plateau and basin.

2 For each of the following features: young fold mountain, block mountain, and rift valley:
 (i) draw a clear diagram to show its main features
 (ii) explain its possible origin
 (iii) name and locate a region where a good example may be found.

3 Explain the meanings of the following terms by using well-labelled diagrams:
 (i) syncline and anticline
 (ii) dip slope and scarp slope
 (iii) tensional force and compressional force.

Objective Exercises

1 Which one of the following groups of terms is applicable to some parts of the ocean floor?
 A basin, deep, cirque, plateau
 B trench, ridge, drumlin, plateau
 C plateau, basin, dune, ridge
 D ridge, deep, basin, waterfall
 E trench, ridge, basin, plateau

 A B C D E
 ☐ ☐ ☐ ☐ ☐

2 Which of the following statements is **not** true in respect of sedimentary rocks?
 A The particles of rock are sometimes completely of organic origin.
 B The rocks are non-crystalline.
 C They are rocks whose structure is determined by great pressure or heat.
 D The rocks have been deposited in layers.
 E The rocks may be formed under water.

 A B C D E
 ☐ ☐ ☐ ☐ ☐

3 Which statement is **not** true of fold mountains?
 A They are often adjacent to a stable area of old crystalline rocks.
 B They contain a core composed of metamorphic and igneous rocks.
 C They form rugged peaks.
 D They are caused by contraction of the Earth's crust.
 E Their tops are often buried beneath snow and ice.

 A B C D E
 ☐ ☐ ☐ ☐ ☐

4 The stratification of sedimentary rocks is mainly caused by
 A lateral pressure being applied from two sides
 B the deposition of rock particles in layers

C the removal by erosion of rocks exposed at the surface
D the compression of rock particles by earth movements

 A B C D
 ☐ ☐ ☐ ☐

5 A rift valley is formed mainly by
 A forces of tension in the Earth's crust
 B the subsidence of the floor of a river valley
 C the formation of fold mountains
 D the over-deepening of a valley by ice action

 A B C D
 ☐ ☐ ☐ ☐

6 Which of the following statements is **irrelevant** when explaining factors involved in the formation of young fold mountains?
 A Thick layers of sediment are crushed between continental blocks.
 B sediments are deposited in a geosyncline.
 C As sediments in the geosyncline accumulate the floor of the geosyncline subsides.
 D Continental blocks move in opposite directions.

 A B C D
 ☐ ☐ ☐ ☐

7 Questions 7 and 8 are based on the map given below which shows some important structural features.
 Which of the following numbers marks the position of an ancient tilted plateau
 A 1
 B 2
 C 3
 D 4
 E 5

 A B C D E
 ☐ ☐ ☐ ☐ ☐

8 A well-developed rift valley is represented by the number
 A 1
 B 2
 C 3
 D 4
 E 5

 A B C D E
 ☐ ☐ ☐ ☐ ☐

3 Earthquakes and Vulcanicity

Destruction caused in an Alaskan city by an earthquake in 1964

EARTHQUAKES

These are sudden earth movements or vibrations in the earth's crust. They are caused by:
(i) The development of faults (cracks) in the crust which result from collision between plates.
(ii) Movements of molten rock below, or within the crust, or the sudden release of stress which has slowly built up along the fault plane.

Nature of Earthquakes

Waves passing out from the point of origin (focus) set up vibrations which may reach up to 200 per minute. The vibrations cause both vertical and lateral movements and this violent shaking causes great destruction to buildings. Earthquakes are frequently associated with faults.

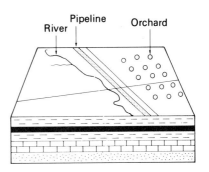

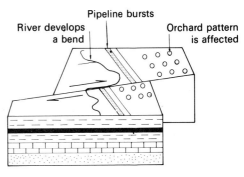

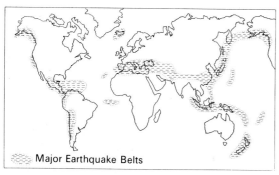

Major Earthquake Belts

Effects of Earthquakes

1 They can cause vertical and lateral displacement of parts of the crust.
2 They can cause the raising or lowering of parts of the sea floor as in Sagami Bay (Japan) in 1923. Parts of the Bay were uplifted by 215 metres (700 feet). This causes *tsunamis* or tidal waves.
3 They can cause the raising or lowering of coastal regions as in Alaska in 1899 when some coastal rocks were uplifted by 16 metres (50 feet).
4 They can cause landslides as in the loess country of North China in 1920 and 1927.
5 They can cause the devastation of cities, fires and diseases.

Some Catastrophic Earthquakes

1755	Portugal	Caused depression of the sea floor near Lisbon
1868	Peru	30 000 people killed
1899	Alaska	Coast of Disenchantment Bay uplifted
1906	California	San Francisco destroyed
1906	Chile	3000 people killed
1920	Japan	Level of Sagami Bay changed and 200 000 people killed
1927	China	Landslides killed 100 000 people
1931	New Zealand	Napier destroyed
1931	Nicaragua	Managua (the capital) destroyed
1960	Agadir (Morocco)	Town destroyed and 10 000 people killed
1962	Iran	Over 20 000 people killed
1970	Peru	Earthquake on 31st May killed 50 000 people and made 1 000 000 people homeless
1972	Nicaragua	Managua, the capital, devasted and 50 000 people killed

VOLCANIC ACTION AND THE FEATURES IT PRODUCES

Rocks below the crust have a very high temperature, but the great pressure upon these keeps them in a semi-solid state. If the pressure weakens (as happens when faulting or folding takes place) then some of the rocks become liquid. This liquid is called *magma*. The magma forces its way into cracks of the crust and may either reach the surface where it forms *volcanoes* or *lava flows*, or it may collect in the crust where it forms *batholiths*, *sills* and *dykes*. Magma reaching the surface may do so quietly or with great violence. Whichever happens it eventually cools and solidifies.

This diagram shows the chief types of volcanic forms. They do not occur together like this.

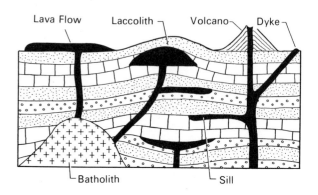

Lava Flow   Laccolith   Volcano   Dyke

Batholith   Sill

Volcanic Features formed in the Crust

BATHOLITH   This is a very large mass of magma which often forms the root of a mountain. It is made of granite and can become exposed on the surface by the removal of the overlying rocks by erosion.

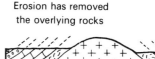

Erosion has removed the overlying rocks

GRANITE

The rocks surrounding the batholith are changed to metamorphic blocks by heat and pressure

SILL   When a sheet of magma lies along the bedding plane it is called a sill. Some sills form ridge-like escarpments when exposed by erosion, e.g. Great Whin Sill in Northern England. Others may give rise to waterfalls and rapids where they are crossed by rivers.

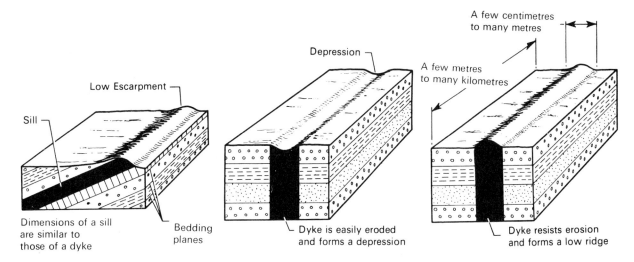

Low Escarpment

Sill

Dimensions of a sill are similar to those of a dyke

Bedding planes

Depression

Dyke is easily eroded and forms a depression

A few centimetres to many metres

A few metres to many kilometres

Dyke resists erosion and forms a low ridge

DYKE When a mass of magma cuts across the bedding planes and forms a wall-like feature it is called a dyke. Dykes may be vertical or inclined. Some dykes when exposed on the surface resist erosion and stand up as ridges or escarpments. Others are easily eroded and form shallow depressions. Examine the diagrams above. Like sills they sometimes give rise to waterfalls or rapids.

Volcanic Features formed on the Surface

Sometimes magma reaches the surface through a *vent* (hole) or a *fissure* (crack) in the surface rocks. When magma emerges on the surface it is called *lava*. If lava emerges via a vent it builds up a *volcano* (cone-shaped mound) and if it emerges via a fissure it builds up a *lava platform* or *lava flow*.

Vent Eruptions and Types of Volcanoes

Structure of a Volcano

The cone is made of either lava, or a mixture of lava and rocks torn from the crust, or ash and cinders (small fragments of lava).

1 Ash and Cinder Cone

When lava is violently ejected it is blown to great heights and it breaks into small fragments. These fall back to earth and build up a cone. Examples: Vulcano de Fuego (Guatemala), Paracutin (Mexico).

2 Lava Cones

The slope of the cones depends upon whether the lava was fluid or viscous when it was molten. Fluid lavas give rise to gently sloping cones, e.g. Mauna Loa (Hawaii).

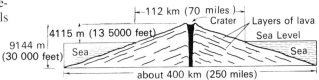

Viscous lavas give rise to steeply sloping cones. Sometimes the lavas are so viscous that when they are forced out of the volcano they form a *spine* or *plug*. Spines are rare because they often rapidly break up on cooling. This was the fate of Mont Pelée.

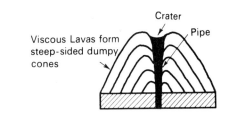

Viscous Lavas form steep-sided dumpy cones

Crater
Pipe

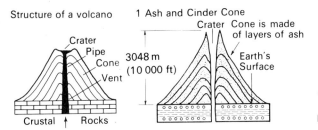

Structure of a volcano

Crater
Pipe
Cone
Vent

Crustal ↑ Rocks

1 Ash and Cinder Cone

3048 m (10 000 ft)

Crater Cone is made of layers of ash

Earth's Surface

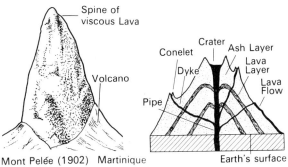

Spine of viscous Lava

Volcano

Mont Pelée (1902) Martinique

Conelet
Dyke
Pipe

Crater Ash Layer
Lava Layer
Lava Flow

Earth's surface

Bromo and Batok in Java

These cones are made of ash and cinders. The sides are deeply grooved by rain erosion. The flat land around the volcanoes consists of fine black volcanic dust.

3 Composite Cone

This is made of alternate layers of lava and ash.

This type of volcano begins each eruption with great violence which accounts for the layers of ash. As the eruption gets under way the violence ceases and the lava pours out forming layers on top of the ash. Lava often escapes from the sides of the the cone where it builds up small conelets. The best examples of this type of volcano are Vesuvius (Italy), Etna (Sicily) and Stromboli (Italy). Sometimes explosive eruptions are so violent that the whole top of the volcano sinks into the magma beneath the vent. A huge crater called a *caldera* later marks the site of the volcano.

Formation of a Caldera

A caldera may become the site of a lake, e.g. Lake Toba (Northern Sumatra) and Crater Lake (U.S.A.). Sometimes eruptions begin again and new cones form in the caldera, e.g. Krakatoa (Sunda Strait). The town of Crater near Aden is built in a caldera.

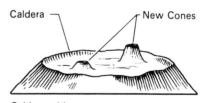

Caldera with new cones

Formation of a Caldera

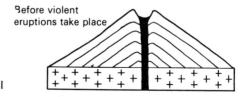

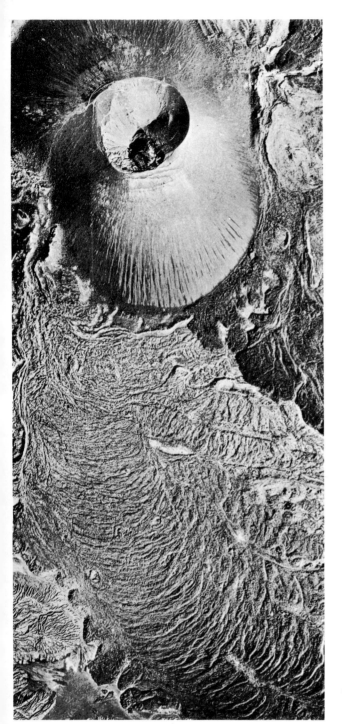

Volcano Cerro de Lopizio in Mexico

In this aerial view the cone and its crater stand out extremely well. This volcano is only 48 kilometres (30 miles) south of Paracutin which first erupted in 1943 and which is now 460 metres (1500 feet) high.

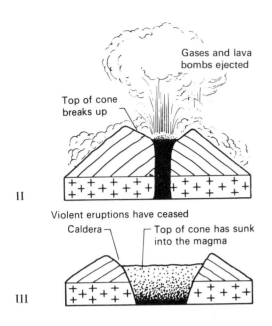

II

III

Mt. Meru in Northern Tanzania

The caldera of this volcano contains a conelet, the vent of which is clearly visible. Note the steep walls of the caldera.

Volcanic Activity

Are Volcanoes always active?

Volcanoes usually pass through three stages in their life cycle. In the beginning eruptions are frequent and the volcano is *active*. Later eruptions become so infrequent that the volcano is said to be *dormant* (sleeping). This is followed by a long period of inactivity. Volcanoes which have not erupted in historic times are said to be *extinct*.

Like all land forms a volcano is attacked by weathering and erosion and by the time it is extinct most if not all the volcano may have been removed. Sometimes the neck of lava remains and this stands up as a steep-sided plug. There are many plugs in central France and some of them became the sites of castles in the Middle Ages (see photograph below).

Fissure Eruptions and the Features they form

Large quantities of lava quietly well up from fissures and spread out over the surrounding countryside. Successive lava flows result in the growth of a lava platform which may be extensive and high enough to be called a plateau. The thickness of such lava plateaus may be as much as 1800 metres (5900 feet) (Deccan lava plateau near Bombay).

Examples of lava plateaus are: Columbia and Snake Plateau of the U.S.A. (520 000 sq km or 325 000 sq miles); north-west Deccan of India (648 000 sq km or 405 000 sq miles); parts of South Africa where the edge of the plateau forms the Drakensberg Mountains; Victoria and Kimberley Districts of Australia.

Rivers crossing lava plateaus often carve out deep gorges, e.g. Snake River of Oregon (North America). Sometimes the lavas weather to give fertile soils, as in the north-west Deccan which is used for cotton cultivation.

Stages in the formation of a Lava Plateau

Original relief

Le Puy in Central France

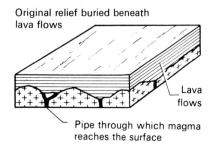

Original relief buried beneath lava flows

Lava flows

Pipe through which magma reaches the surface

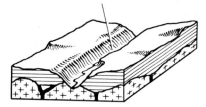

River valley cut through lava into rocks below

Stages in the formation of a lava plateau

Why do some volcanoes erupt violently?

Most magma contains gases which are under great pressure. In some instances there is a sudden decrease in pressure in the rising magma which causes the gases in it to expand very rapidly. This sudden expansion can cause violent explosions. Water vapour is often one of the gases and it may have originated from the magma or from water in the crust with which the magma came into contact. Many of the gases burn with a fierce heat and some of them like sulphur gases can form a dense cloud which rolls down the side of the volcano killing everything in its path. When eruptions are violent the lava explodes into small pieces which are blown to great heights. The sizes of the pieces vary from grains to small chunks of rock. The latter are called volcanic bombs. If the explosions are particularly violent the fine dust can reach such great heights that it gets carried along by the air currents of the upper atmosphere. When Krakatoa exploded in 1883 some of its dust passed right round the world causing vivid sunsets in many countries.

Other forms of Volcanic Activity

Emissions of gases and steam periodically take place from dormant volcanoes. Similar emissions of gases and steam take place in some volcanic regions where active lava eruptions have long since ceased. Superheated water may flow quietly as in *hot springs*, or it may be thrown out with great force and accom-

panied by steam as in *geysers*. Thus a geyser differs from a hot spring in that its water is ejected explosively. Geysers often form *natural fountains*. Hot springs and geysers are common in Iceland, North Island of New Zealand and the Yellowstone National Park of U.S.A.

Formation of a Geyser

1 *Based on underground cavern*

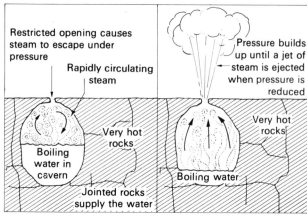

Restricted opening causes steam to escape under pressure

Rapidly circulating steam

Very hot rocks

Boiling water in cavern

Jointed rocks supply the water

Dormant

Pressure builds up until a jet of steam is ejected when pressure is reduced

Very hot rocks

Boiling water

Ejecting steam

When the jet of steam has been ejected the geyser returns to the dormant phase. These geysers are often very regular, e.g. *Old Faithful* in the Yellowstone National Park of the U.S.A., ejects steam every half an hour.

2 *Based on underground cavern and sump*

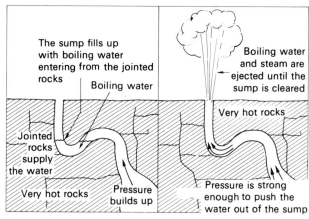

The sump fills up with boiling water entering from the jointed rocks

Boiling water

Jointed rocks supply the water

Very hot rocks

Pressure builds up

Dormant

Boiling water and steam are ejected until the sump is cleared

Very hot rocks

Pressure is strong enough to push the water out of the sump

Ejecting water and steam

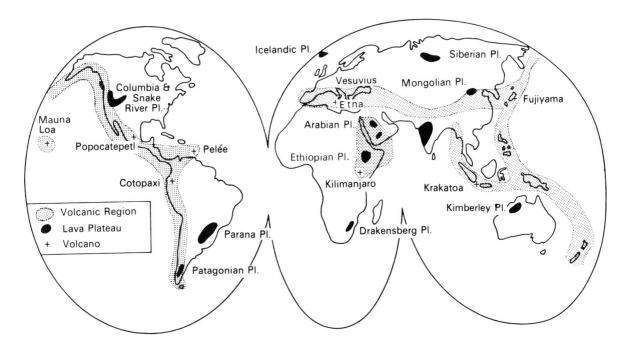

World Distribution of Volcanoes and Lava Plateaus

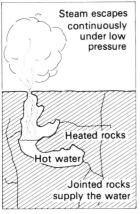

Formation of a Fumerole

Steam escapes continuously under low pressure

Heated rocks

Hot water

Jointed rocks supply the water

If the steam contains much dissolved sulphurous gas, the vent becomes surrounded by yellow sulphur deposits. This type is known as a *solfatara*.

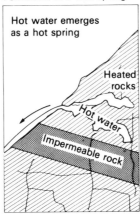

Formation of a Hot Spring

Hot water emerges as a hot spring

Heated rocks

Hot water

Impermeable rock

The hot water contains minerals dissolved from the rocks. The water is believed to be good for rheumatism and similar ailments which has resulted in Spas (health resorts) developing at many hot springs.

Influence of Volcanic Eruptions on Man

Destructive Influences

1 Some eruptions cause great loss of life, e.g. Vesuvius in A.D. 79 (out-pourings of gases and ash); Krakatoa in 1883 (caused great sea waves which drowned 40 000 people in neighbouring islands); Mont Pelée in 1902 (out-pourings of gases killed 30 000 people).

2 Some eruptions cause great damage to property, e.g. Vesuvius buried Herculaneum and Pompeii with ash; Mont Pelée caused the destruction of St. Pierre.

Constructive Influences

1 Some lava out-pourings have weathered to give fertile soils, e.g. in Java, the north-western part of the Deccan Plateau, and the plains around Etna. These regions are of important agricultural value.

2 Volcanic activity sometimes results in the formation of precious stones and minerals. These occur in some igneous and metamorphic rocks, e.g. diamonds of Kimberley; copper deposits of Butte (U.S.A.); and the nickel deposits of Sudbury in Canada.

3 Some hot springs are utilised for heating and supplying hot water to buildings in New Zealand and Iceland.

37

EXERCISES

1 Explain the main differences in appearance and origin between the members of each of the following pairs of features:
   (i) batholith and lava flow;
   (ii) sill and dyke;
   (iii) crater and caldera;
   (iv) hot spring and geyser.
   Illustrate your answer with diagrams.

2 Carefully explain the differences between the appearance and the origin of (i) a lava volcano, and (ii) a composite volcano. Draw a large diagram for each type of volcano and on this mark and name the main parts.

3 Locate by shading on a map of the world, *two* important volcanic and earthquake regions. Briefly describe one major volcanic eruption and one major earthquake which have occurred in the last twenty years. Your description should indicate the causes and effects of these natural catastrophes.

4 (a) Carefully explain the differences between:
       (i) igneous and sedimentary rocks;
       (ii) sedimentary and metamorphic rocks;
       (iii) sial and sima.
   (b) Carefully explain how a sedimentary rock originates and name two common examples of this rock.

Objective Exercises

1 Which one of the following best describes the world distribution of active and recently active volcanoes?
   A they are found in association with young fold mountain chains
   B they occur in river flood plains
   C they are associated with old eroded mountain chains
   D they are located on the western sides of continents
   E they tend to form chains around ocean basins

   A B C D E
   ☐ ☐ ☐ ☐ ☐

2 Which of the following features may occur when lava cools at the surface?
   A basalt plateau
   B sill
   C batholith
   D dyke

   A B C D
   ☐ ☐ ☐ ☐

3 An intrusion of magma along a bedding plane is called a
   A dyke
   B sill
   C batholith
   D laccolith

   A B C D
   ☐ ☐ ☐ ☐

4 Which of the following features is the product of vulcanicity?
   A geosyncline
   B escarpment
   C atoll
   D fold mountain
   E caldera

   A B C D E
   ☐ ☐ ☐ ☐ ☐

5 A volcanic eruption is most likely to be violent when
   A the volcano is near to the sea
   B the neck of the volcano is sealed by a plug
   C the lava is viscous
   D the lava reaches the surface through a fissure
   E the volcano is on the ocean floor

   A B C D E
   ☐ ☐ ☐ ☐ ☐

6 Which of the following features is **not** an aspect of vulcanicity?
   A geyser
   B batholith
   C dyke
   D fold

   A B C D
   ☐ ☐ ☐ ☐

7 Which of the following statements best describes the usefulness of vulcanicity?
   A A violent volcanic eruption represents the release of an enormous amount of energy.
   B The high density of population in parts of Java is dependent on agriculture.
   C Some volcanoes are dormant.
   D Fissure eruptions are usually non-violent.

   A B C D
   ☐ ☐ ☐ ☐

4   Features Produced by Running Water

External forces lower the level of the land by wearing it away and this process is called *denudation*. They also raise the level of the land by *deposition*. Denudation consists of (i) Weathering, (ii) Erosion. *Weathering* refers to the gradual disintegration of rocks which lie exposed to the weather. The effect of weathering can be seen on stone monuments and buildings where pieces of stone have flaked off, and on iron railings which have rusted.

Erosion refers to the disintegration of rocks which lie exposed to what are called the agents of erosion, i.e. running water, wind and moving ice.

Deposition refers to the laying down of rock particles by the agents of erosion.

1.   WEATHERING AND THE FEATURES IT PRODUCES

It is effected by physical forces and chemical forces.

Physical Weathering
I By Temperature Changes

In arid regions, such as hot deserts, rock surfaces heat up rapidly when exposed to the sun and the surface layers expand and break away. At night when the temperature falls rapidly the same layers contract and more cracks develop. In time the layers of rock peel off and fall to the ground. Rock break-up of this type is called *exfoliation*. Exfoliation is best seen in rocks of uniform structure. This process ultimately changes rocky masses into rounded boulders.

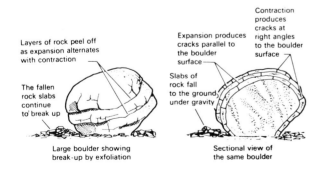

Layers of rock peel off as expansion alternates with contraction

The fallen rock slabs continue to break up

Expansion produces cracks parallel to the boulder surface

Contraction produces cracks at right angles to the boulder surface

Slabs of rock fall to the ground under gravity

Large boulder showing break-up by exfoliation

Sectional view of the same boulder

Exfoliation domes are common in the Kalahari, Egyptian and Sinai Deserts. Physical weathering on steep slopes often produces *screes* which collect at the bottom of the slopes.

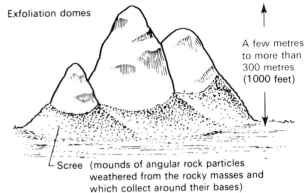

Exfoliation domes

A few metres to more than 300 metres (1000 feet)

Scree (mounds of angular rock particles weathered from the rocky masses and which collect around their bases)

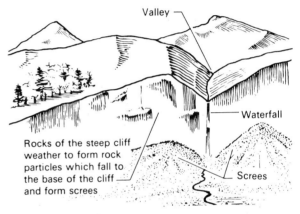

Valley

Waterfall

Rocks of the steep cliff weather to form rock particles which fall to the base of the cliff and form screes

Screes

II By Frost Action

When water freezes its volume increases. If water in the cracks of rocks freezes a tremendous power is applied to the sides of the cracks and the cracks widen and deepen.

Frost action in time breaks up rocky outcrops into angular blocks which later break up into small fragments.

Frost action is very common in the winter season in temperate regions and in some high mountains all the year, e.g. the Himalayas. It usually involves the freezing of water in the cracks of rocks during the night and the thawing of the ice during the following day. The angular fragments of rocks which break off the main rocky masses through frost action form screes around the lower slopes of the rocky outcrops.

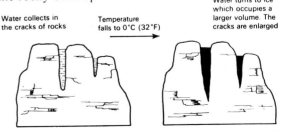

Water collects in the cracks of rocks

Temperature falls to 0°C (32°F)

Water turns to ice which occupies a larger volume. The cracks are enlarged

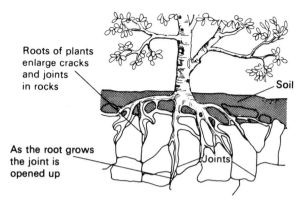

Roots of plants enlarge cracks and joints in rocks

Soil

As the root grows the joint is opened up

Joints

Desert Screes near St. Catherine's Monastery in Sinai

Note Screes formed by frost action contain *angular* rock particles, those formed by other types of weathering contain *rounded* rock particles.

Some rocks break up into large rectangular-shaped blocks under the action of mechanical weathering. This may be partly frost action and partly expansion and contraction through temperature changes. This is called *block disintegration.*

Chemical Weathering
I By Rain Action

Rain is really a weak acid because it dissolves oxygen and carbon dioxide as it falls through the air. Some minerals especially carbonates are dissolved out of rocks by rainwater. These rocks are therefore weakened and begin to break up. Chemical weathering is most active in limestone rocks the surface of which become weathered into deep narrow grooves called *grikes* which are separated by flat or round-topped ridges called *clints.* The rainwater enters the limestone via the joints which become enlarged to form clints.

In humid tropical countries chemical weathering is very active in many types of rocks.

Joints are opened by both frost action and expansion and contraction

Block disintegration

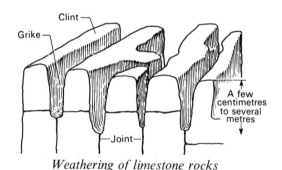

Clint

Grike

A few centimetres to several metres

Joint

Weathering of limestone rocks

III By Plant and Animal Action

The roots of plants, especially trees, can force joints and cracks apart in rocks. Some animals by burrowing also help to break up rocks.

Note A covering of vegetation often protects rocks from weathering. It binds the soil together and reduces changes in temperature. The removal of a covering of vegetation can result in *soil erosion.*

II By Plants and Animals

Bacteria in the presence of water break down certain minerals in the soil, and all plants absorb minerals from the soil. Decaying vegetation produces organic acids which cause a further break-down of minerals. All these actions help to weaken and break up the rocks.

Note Chemical weathering takes place in all regions where there is rain, but it is most marked in wet regions which have high temperatures. Physical weathering takes place in all regions where there are changes in temperature but it is most marked in the hot deserts which have a large daily temperature range.

2. UNDERGROUND WATER AND THE FEATURES IT PRODUCES

When rain falls some of it runs off the surface forming streams and rivers; some of it evaporates directly, or indirectly via plants; some of it soaks into the surface rocks.

The amount of run-off, evaporation and percolation depends upon the nature of the rocks, the slope of the land and the climate. Run-off on steep slopes is greater than on gentle slopes; evaporation in dry climates is greater than in humid climates, and water percolates into sands more easily than into granites.

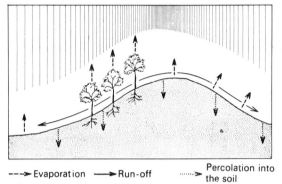

---> Evaporation ——> Run-off ·····> Percolation into the soil

How Water enters the Rocks

Water will enter rocks which are *porous* (i.e. rocks having small air spaces, e.g. sandstone) or rocks which are *pervious* (i.e. rocks having joints or cracks, e.g. granite). Rocks which allow water to pass through them are said to be *permeable*. Sandstone is a permeable rock. Rocks which do not allow water to pass through them are said to be *impermeable*. Clay is an impermeable rock.

Note Clay is porous (water enters it), but it is impermeable (water will not pass through it).

The Water-table or the Level of Saturation

Water entering the surface rocks moves downward until it comes to a layer of impermeable rock when further downward movement ceases. There are three water zones below the surface:

(i) *The zone of permanent saturation.*
    The pores of the rocks of this zone are always filled with water.

(ii) *The zone of intermittent saturation.*
    The pores of the rocks of this zone contain water only after heavy rain.

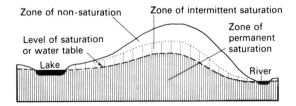

(iii) *The zone of non-saturation.*
    This lies immediately below the surface. Water passes through but never remains in the pores of the rocks of this zone.

Springs

When water flows naturally out of the ground it is called a *spring*. There are many types of springs. Here are some of the more common ones.

I A permeable rock lying on top of an impermeable rock in a hill.

This diagram shows two lines of springs which occur where the junction of the two rock layers meets the surface. Notice that one line of springs is temporary.

I   Intermittent spring occurring in wet season only
P   Permanent springs
– – Water Table during the wet season
····· Water Table during the dry season which is too low to supply the intermittent springs

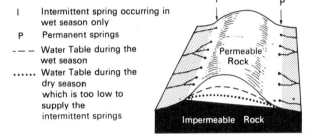

II Well-jointed rocks forming hilly country produce springs.

Water enters the rocks via the joints. Springs frequently occur where the water-table meets the surface.

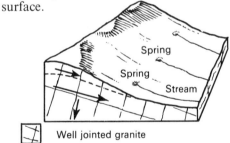

    Well jointed granite
– – – Water table
——> Movement of water in the rock

III The impounding of water by a dyke can give rise to springs.

If a dyke cuts across a layer of permeable rock then the water on the up-slope side of the dyke is impounded. The water-table here rises and it gives rise to springs where it meets the surface.

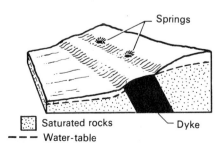

    Saturated rocks
– – – Water-table

IV Chalk or limestone escarpments which overlie impermeable rocks give rise to springs.

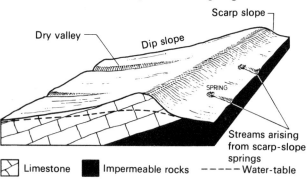

Dry valley — Dip slope — Scarp slope — SPRING — Streams arising from scarp-slope springs

▨ Limestone ■ Impermeable rocks ----- Water-table

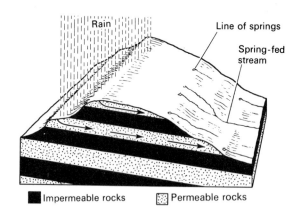

Rain — Line of springs — Spring-fed stream

■ Impermeable rocks ▨ Permeable rocks

The dip slope with the scarp slope (sometimes called escarpment) together form a feature known as a *cuesta*. On the dip slope of cuestas such as the North Downs in Southern England, there are many dry valleys along which streams once flowed. These valleys are now dry because the level of the water-table has fallen. The valleys are called *dry valleys*. After a prolonged period of heavy rain the water-table sometimes rises high enough to allow a seasonal flow of water. This is called a *bourne* and an example of where it occurs is in the Caterham Valley near Croydon.

Usually two lines of springs occur, one at the foot of the scarp slope and the other on the dip slope. Since there is little or no surface drainage in limestone or chalk regions, settlements often become located near to the springs.

V Gently sloping alternate layers of permeable and impermeable rocks often produce springs.
Rain falling on the exposed ends of the permeable rock layers soaks down the sloping bedding planes and finally comes out as springs. The springs are sometimes in lines.

VI Numerous springs occur where the junction of limestone rocks and underlying impermeable rocks meets the surface.
Limestone regions rarely have surface drainage.

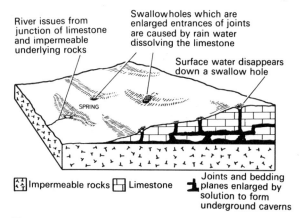

River issues from junction of limestone and impermeable underlying rocks

Swallowholes which are enlarged entrances of joints are caused by rain water dissolving the limestone

Surface water disappears down a swallow hole

SPRING

▨ Impermeable rocks ▨ Limestone ▸ Joints and bedding planes enlarged by solution to form underground caverns

Both the rain and streams entering the region work their way by solution into the limestone rocks. Some joints become enlarged to form *swallow holes*. If the limestone rocks rest on impermeable rocks then this water will reach the surface again where the two rock types meet the surface. The water may issue out as streams or springs. Inside the limestone, underground streams dissolve the rocks and form huge caverns.

Wells
A well is a hole sunk in the ground to below the water-table. Water then seeps out of the rocks into the well.

Wells which are sunk far below the water-table always contain water (A, B and C). Wells sunk only just below the water-table often go dry in periods of drought when the water-table falls (D). If a well does not reach the water-table it will be dry.

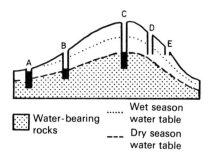

▨ Water-bearing rocks ..... Wet season water table --- Dry season water table

Artesian Basins and Artesian Wells
Artesian Basin
An artesian basin consists of a layer of permeable rock lying between two layers of impermeable rock such that the whole forms a shallow syncline with one or both ends of the permeable rock layer exposed on the surface. Rain water enters the permeable layer at its exposed ends. This layer becomes saturated with water and is called an *aquifer* Western Australia, the Sahara Desert and parts of North America from Saskatchewan to Kansas are underlain by extensive artesian basins.

The diagram on the right shows a part of the artesian basin of the Sahara Desert. In places the aquifer bends up towards the surface and wind erosion sometimes exposes it. When this happens pools of water occur and these are called *oases* (sing. *oasis*). If the aquifer is near to the surface wells can be sunk. This is often done. Notice some typical hot desert erosional and depositional features, e.g. rock platforms, wadis and sand dunes. Also notice that the exposed part of the aquifer which receives the rain is called the *catchment area*. The London Basin consists of a shallow syncline formed of chalk which lies between layers of clay. In some parts the water-table has fallen by as much as 30 metres (100 feet) during the last 50 years. One reason for this is that Waterboards and industries have taken so much water from the numerous wells sunk in the chalk aquifer.

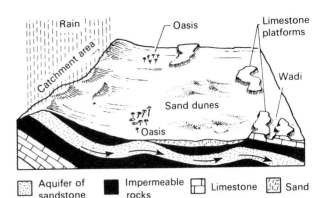

Section across part of the Sahara Desert

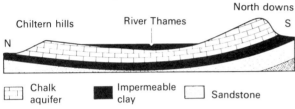

Section across the London Basin

Artesian Well

If a well is sunk in the aquifer of an artesian basin and the pressure of water is sufficient to cause the water to flow out of it, then the well is called an *artesian well*.

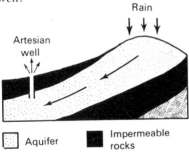

Value of Ground Water to Man

1 Springs and wells have played an important part in the siting of settlements in many regions all over the world.
2 In some regions, e.g. hot deserts and semi-arid plains, settlement is only possible by utilising the ground water. When this is too deep to be tapped, settlement does not take place.
3 In southern Algeria there is a limestone plateau which is 600 metres (1950 feet) above sea level. The Chebka people who live there have dug wells to tap the underground water. The water is used for irrigating the land and the wells have given rise to numerous oases.
4 The Soafas (people of the Suf) who live on the

borders of Tunis and Algeria use artesian water which lies near to the surface. This water maintains many oases which cultivate date palms.
5 The great artesian basins which underlie Queensland, New South Wales and parts of South Australia have an area of 1 500 000 sq km (58 600 sq miles). Water taken from the many wells which tap the aquifer of the basins is too salty for irrigation but is used for watering large herds of cattle.
6 Similar artesian basins underlie parts of North America from Kansas to Saskatchewan and wells here raise water for use in cattle ranching.
7 Wells play an important part in agriculture in the Indo-Gangetic Plain of the Indian sub-continent in the dry season. The water is used for irrigation.

3. RAIN ACTION AND THE FEATURES IT PRODUCES.

Rain action is an aspect of erosion because it involves movement. It is most marked in semi-arid regions because these have little or no vegetation and the rains, though infrequent, are torrential. Rain action produces many types of features of which the following are the most common: *gully, earth pillar, soil creep,* and *landslide*.

Gully

Rain which falls on gently sloping land which has little or no vegetation moves downhill as a sheet of water. The slope is quickly eroded into deep grooves called *gullies*. They develop on a small scale on embankments and cuttings and also tip heaps. They develop on a large scale in semi-arid regions where the landscape becomes cut up into gullies and ridges of all shapes and sizes, as in the Badlands of the Dakotas (U.S.A.). See page 44.
The influence of rain-wash on soil erosion will be discussed in a later section.

Earth Pillar

When rain falls on slopes made of clay and boulders, the clay is rapidly removed except where boulders

43

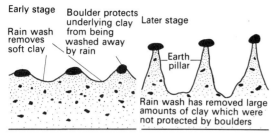

Early stage

Rain wash removes soft clay

Boulder protects underlying clay from being washed away by rain

Later stage

Earth pillar

Rain wash has removed large amounts of clay which were not protected by boulders

Note   Earth pillars range from a few centimetres to several metres in height.

form a protection. When this happens, columns of clay, capped by boulders, develop. These are called *earth pillars*. However, the pillars are only temporary. In time they will be removed by erosion.

Soil Creep (slow movement)

On all sloping land there is a steady movement of the soil down the slope. Rain water soaking into the soil slowly trickles down the slope and this causes a movement in the soil. Bulging fences and walls and outward bending tree trunks reflect this movement which is called *soil creep*.

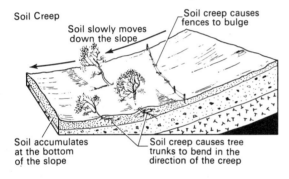

Soil Creep

Soil creep causes fences to bulge

Soil slowly moves down the slope

Soil accumulates at the bottom of the slope

Soil creep causes tree trunks to bend in the direction of the creep

Badlands of Dakota in U.S.A. The hill sides are deeply gullied

Earth Pillars in Upper Rhône Basin

Gully Erosion in a road-side laterite bank in Kuala Lumpur

Landslide (sudden movement)

A *landslide* takes place when large quantities of loosened surface rocks slide down steep slopes which may be cliff faces, embankments, valley sides or railway cuttings. Landslides are caused by the lubricating action of rain water and the pull of gravity which result in slumping or sliding.

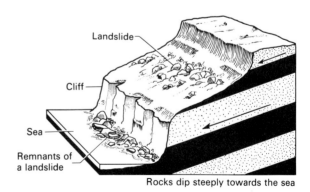

Landslide

Cliff

Sea

Remnants of a landslide

Rocks dip steeply towards the sea

Landslide caused by sliding   The black arrows show both the place of lubrication and the direction of movement.

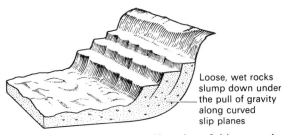

Loose, wet rocks slump down under the pull of gravity along curved slip planes

Landslide caused by slumping Slumping of this type takes place on steep slopes made of clay. It is especially common in cliffs of clay which are under wave attack.

Landslide on a Railway Cutting

Landslides are very common on the sides of railway and road cuttings, especially in mountainous regions which have heavy falls of rain. In some regions frost action speeds up the process. Frozen soils and subsoils on steep slopes become unstable when they thaw and the movement of water down the slope together with the pull of gravity triggers off a landslide.

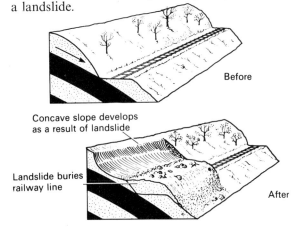

Before

Concave slope develops as a result of landslide

Landslide buries railway line

After

4. RIVER ACTION AND THE FEATURES IT PRODUCES

A river's *source* is the place at which it begins to flow. It may be in the melt waters of a glacier, e.g. the Rhône (France), or in a lake, e.g. the Nile (Africa), or in a spring, e.g. the Thames (England), or in a region of steady rainfall, e.g. the Zaire (Africa). A river's *mouth* is the place where the river ends. This is usually in the sea, e.g. the Amazon (Atlantic), the Niger (Gulf of Guinea) and the Indus (Arabian Sea), although it may be in a lake, e.g. the Volga (Caspian), or in a salt swamp, e.g. the Chari River (Lake Chad) and the Tarim River (Lop Nor).

Rivers are one of the greatest sculpturing agents at work in humid regions. They carve out valleys in the highlands and as they do so they produce peaks, ridges and hills.

The material so removed is transported from the highlands and is deposited around them as gently sloping plains. A river thus does three types of work: it *erodes*; it *transports*; and it *deposits*.

In process of time river erosion, transport and deposition turn the original surface into an almost level plain which is called a *peneplain*.

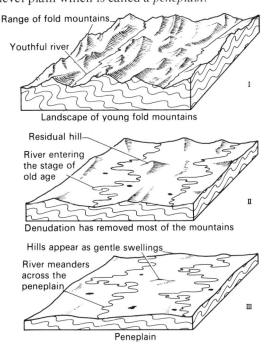

Range of fold mountains

Youthful river

I

Landscape of young fold mountains

Residual hill

River entering the stage of old age

II

Denudation has removed most of the mountains

Hills appear as gentle swellings

River meanders across the peneplain

III

Peneplain

The River as an Energy System

Energy is the ability to do work, and the amount of energy which a river has determines whether it can effectively erode its valley and transport the material it is carrying, or whether it drops the material in the form of deposition. We often say that a mature river has three sections: the upper course where most of the erosion takes place, the middle course where transport is the dominant process, and the lower course where the main process is deposition. This is a very simplified view to take as we shall now see. The amount of erosion that a river can achieve depends on its energy. A river's energy increases with its volume and with its velocity, and with its regime, that is, seasonal flow. This means that a large, fast-flowing river will have more power to effect erosion in time of flood than the same river will have in time of drought when it flows sluggishly with little water in its channel. However, not all the energy in a river is used for erosion. Some is needed to overcome frictional resistance both externally along the bed and banks and internally when currents, caused by turbulence, splash and roll against each other. Energy is also needed to transport the pebbles, sand, silt and the dissolved minerals in the water. The following diagram summarises these different uses of energy.

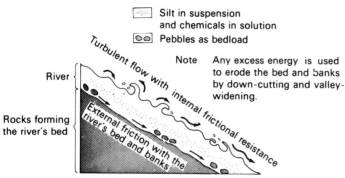

- ▦ Silt in suspension and chemicals in solution
- ◎◎ Pebbles as bedload

River

Rocks forming the river's bed

Turbulent flow with internal frictional resistance

External friction with the river's bed and banks

Note Any excess energy is used to erode the bed and banks by down-cutting and valley-widening.

Long Section of a steeply-flowing River

Energy is used to overcome friction

The shape of a river's channel is also important in determining how much energy a river will have for erosion. A flat, wide channel is very inefficient for transporting water, while a narrow, deep channel is much more efficient because the smaller surface area results in less frictional drag. The diagram below shows two river channels which have the same cross-sectional area and which therefore can carry the same volume of water. But the wide channel has a greater surface area of bed and banks, and therefore frictional drag will be greater than in the narrower channel.

In the upper part of a river's course the gradient of the channel is steeper but the volume of water is less. It used to be thought that a river flowed fastest at this point. This is true, in that the maximum velocity occurs here, but the water splashes and eddies so much that the average speed of the river at this point is not very great. In the middle and lower sections of a river's course the average flow of the river is at least equal to, and in many cases, greater than the speed in the upper course. This is because of the greater loss of energy to overcome internal friction in the turbulent upper section.

Cross Section of an inefficient channel shape

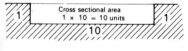

Cross sectional area
1 × 10 = 10 units

Total distance across the bed and banks is 12 units and this river has to overcome more friction and it therefore has less energy for erosion

Diagram to explain the influence of the shape of a river's channel

Cross Section of an efficient channel shape

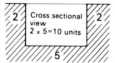

Cross sectional view
2 × 5 = 10 units

Total distance across the bed and banks is 9 units and this river has to overcome less friction and it therefore has more energy for erosion

Energy is used to transport sediments

A river transports its load in three ways: by *traction* or the dragging of the *bedload* of pebbles along its

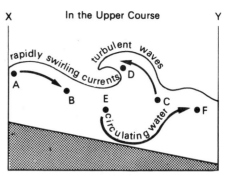

In the Upper Course

rapidly swirling currents

turbulent waves

circulating water

Greater maximum velocity

An object is tracked in its movement from X to Y and the time is recorded. It follows an indirect path (A to B, C to D and E to F) which is greater than the direct path. The distance travelled in the time recorded is greater than it is in the Middle Course.

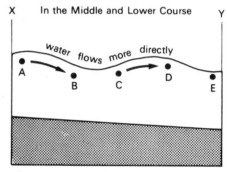

In the Middle and Lower Course

water flows more directly

Greater average velocity

The object is tracked in its movement from X to Y and the time is recorded. It follows a more direct path (A to B, C to D and E to F) but it takes about the same time to do so. This shows that the velocity in the Middle Course is lower than it is in the Upper Course

Diagram to explain the effects of turbulent flow

bed, by *suspension* of light sediments such as silt and mud in the water, and by *solution* of chemicals which are dissolved in the water. Sediments are transported by a river until it has insufficient energy to move them farther. It then deposits them. A river may lose energy where there is either a decrease in gradient, or a widening or meandering of its channel, e.g. in its lower course, or where there is a decrease in volume, e.g. after a flood.

Energy used in erosion

Erosion in a river is caused by *attrition, corrasion, hydraulic action* and *chemical solution*. Attrition is the process whereby pebbles are eroded by striking together as they are rolled along a river's bed. Corrasion is the wearing away of the bed and the banks, by a river's load. Hydraulic action is the wearing away of the bed and the banks of a river by the sheer weight of water hurled against them. This is particularly effective in fast-flowing rivers. Chemical solution is the dissolving of minerals from the rocks, and it is particularly effective with soluble minerals.

In parts of a river's course the main action of the river is erosional, and in these parts the river's surplus energy keeps the channel clear of debris and sediment. In other parts of a river's course, deposition is dominant. This happens when most of a river's energy is used up to overcome frictional drag. The channel then becomes choked with boulders, or banks of sand and shingle if the river has a large load. Between these two, there is a part of a river's course where there is a balance between erosion and deposition, where both processes occur but without one being dominant. The erosional areas more commonly occur in the upper parts of a river's course and it is here that a river appears to have more energy. The depositional areas more commonly occur near to a river's mouth, and it is here that a river has little or no surplus energy for erosion. Based on these general observations a river valley can be divided into an upper section, a middle section and a lower section, and by studying each of these we can see what landscape features are produced in each of the three parts.

A river's volume decreases:
1 When it enters an arid region (especially a hot one)
2 When it crosses a region composed of porous rocks, e.g. sand and limestone
3 In the dry season or in a period of drought.

A river's speed decreases when it:
1 Enters a lake

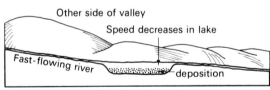

Section along a river valley

2 Enters the sea

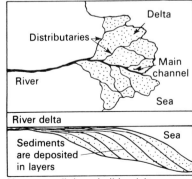

Note: not all rivers build a delta when they enter the sea

3 Enters a flat or gently sloping plain such as a valley bottom.

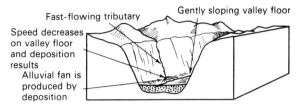

THE DEVELOPMENT OF A RIVER VALLEY

The force of a river partly depends upon its size and partly upon its *gradient*, that is the distance it has to fall before it reaches *base level*. Base level is the surface of a lake or a river or the sea into which it flows. The level of a river at its confluence with a tributary is the base level for that tributary. The upper part of a river, because it is high above the base level, will be able to deepen its channel rapidly. A river cannot, of course, erode below its base level. A river, like an animal or a plant, has a *life-cycle*. In the beginning, when it is in the *stage of youth*, it flows turbulently in a narrow, steep-sided valley whose floor is broken by pot holes and waterfalls. As time passes denudation widens the valley and lowers its floor. Now that the gradient is reduced the river has less energy to erode and the initial bends that it had, because of the nature of its valley floor, become more pronounced. It is now in the *stage of maturity*. As denudation continues the valley is opened out more and more. The gradient is further reduced and deposition now becomes active. Layers of sediments are dropped by the river and these ultimately extend over the entire floor of its valley where they build up a gently sloping plain called a *flood plain*. The river wanders in great *meanders* or loops across this plain and often it becomes divided into many channels by its own deposition. The river is now in the *stage of old age*. Deposition within the mouth of an old river sometimes builds up a triangular-shaped piece of land called a *delta*.

Many river valleys such as those of the Nile, Indus and Irrawaddy contain all three stages. The torrent or *upper course* represents the stage of youth; the valley or *middle course* represents the stage of maturity, and the plain or *lower course* represents the stage of old age.

Note Youthful Stage is often called Torrent Stage. Mature Stage is often called Valley Stage. Old Age Stage is often called Plain Stage.

A river valley grows in length by *headward erosion*. Rain wash, soil creep and undercutting at the head of a river combine to extend the valley up the slope. A river's valley is deepened by *vertical erosion* and widened by *lateral erosion*. The former is entirely a river process; the latter is effected by weathering on the valley sides and by the river on the river banks. When a river's gradient is steep, i.e. when it is in the stage of youth, vertical erosion is dominant. When a river's gradient is very gentle, i.e. when it is in the stage of old age, there is little erosion and deposition is dominant. In a mature valley lateral erosion is dominant.

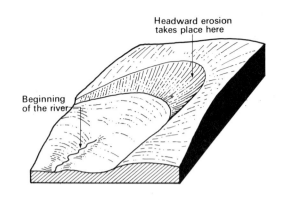

Headward erosion takes place here

Beginning of the river

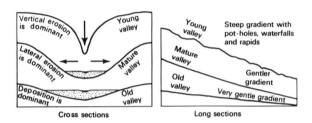

Cross sections

Long sections

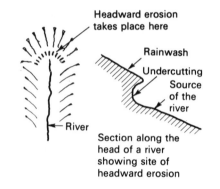

Headward erosion takes place here

Rainwash

Undercutting

Source of the river

River

Section along the head of a river showing site of headward erosion

Over 1 in 10 Youthful (Torrent) Stage.
Between 1 in 10 and 1 in 100 Mature (Valley) Stage.
Under 1 in 100 Old Age (Plain) Stage.

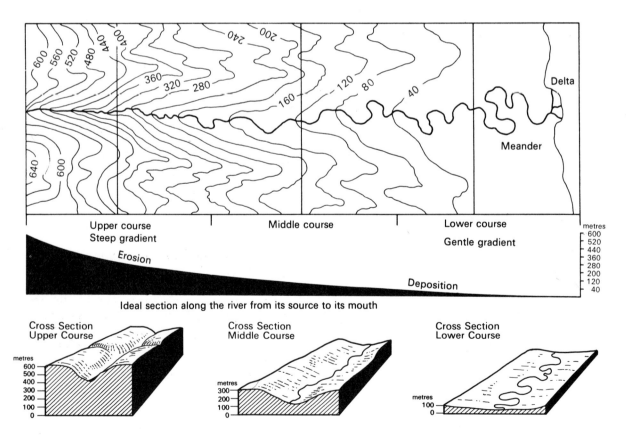

Ideal section along the river from its source to its mouth

Cross Section Upper Course

Cross Section Middle Course

Cross Section Lower Course

THE CHARACTERISTIC FEATURES OF A YOUTHFUL VALLEY

1 Deep, narrow valley (V-shaped)
2 Valley has a steep gradient
   (river is fast-flowing)
3 Pot-holes
4 Interlocking spurs
5 Waterfalls and rapids

Pot-holes

The water of a fast-flowing river swirls if the bed is uneven. The pebbles carried by a swirling river cut circular depressions in the river bed. These gradually deepen and are called *pot-holes*. Much larger but similar depressions form at the base of a waterfall. These are called *plunge pools*.

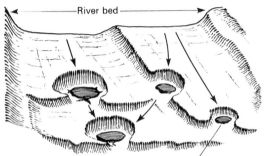

Swirling water falling into a slight depression turns it into a cylindrical hole called a pot hole

Pot hole in a river's bed

Interlocking Spurs

Vertical erosion rapidly deepens the valley. The river twists and turns around obstacles of hard rock. Erosion is pronounced on the concave banks of the bends and this ultimately causes spurs which alternate on each side of the river to *interlock*. The undercut concave banks often stand up as *river cliffs*. On the opposite convex banks there is little or no erosion. The banks form gentle *slip-off slopes*.

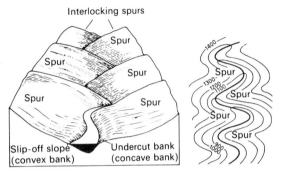

V-shaped valley with interlocking spurs

Waterfalls and Rapids

These occur where the bed of a river becomes suddenly steepened. Waterfalls are of two types:

(i) Those caused by differences in rock hardness into which the river is cutting.
(ii) Those caused by uplift of the land, lava flows and landslides, etc.

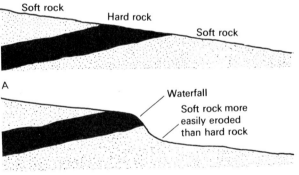

Waterfalls caused by Differences in Rock Hardness

When a layer of hard rock (rock which resists erosion) lies across a river's course, the soft rocks on the down-stream side are more quickly eaten into than is the hard rock. The river bed is thus steepened where it crosses the hard rock and a *waterfall* or a *rapid* develops. A waterfall arises when the hard rock layer is: (i) horizontal, (ii) dips gently up-stream, or (iii) is vertical.

(i) Rock layer is horizontal

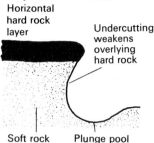

Horizontal hard rock layer

Undercutting weakens overlying hard rock

Soft rock   Plunge pool

(ii) Rock layer dips up-stream

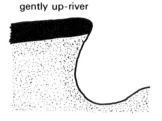

Hard rock layer dips gently up-river

(iii) Rock layer is vertical

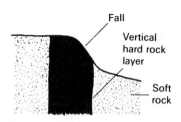

Fall

Vertical hard rock layer

Soft rock

A rapid develops when:
1 A waterfall of types (i) and (ii) above retreats up-stream.
2 A hard rock layer dips down-stream.

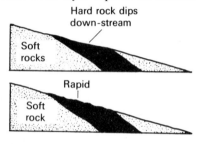

Hard rock dips down-stream

Soft rocks

Rapid

Soft rock

Some Good Examples of Waterfalls and Rapids

1 *Gersoppa Falls* (253 metres) (827 feet) — Western Ghats of India (in the wet season it is the greatest fall in the world).

2 *Kaieteur Falls* (225 metres) (736 feet) — Potaro River (British Guiana): 5° 0′N; 59° 0′W.

3 *Aughrabies Falls* (137 metres) (448 feet) — Orange River (South Africa): 28° 49′S; 20° 22′E.

4 *Victoria Falls* (110 metres) (360 feet) — 49′S; 25° 51′E. The gorge below the Falls is 96 kilometres long.

5 *Niagara Falls* (52 metres) (170 feet) — Between Lakes Erie and Ontario: 43° 7′N; 79° 1′W. The gorge below the Falls is 12 kilometres long.

6 *Livingstone Falls* (274 metres) (900 feet) — Zaire River: 5° 0′S; 14° 15′E. The Falls are formed by 32 rapids.

7 *Nile Cataracts* — Between Aswan and Khartoum.

Young Valleys of Special Interest

Some valleys have very steep sides and are both narrow and deep. These are called *gorges*. A gorge often forms when a waterfall retreats up-stream. The diagrams below show how this takes place. One of the most impressive gorges formed in this way lies below the Victoria Falls.

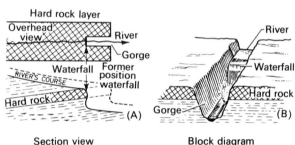

Hard rock layer

Overhead view

River

Gorge

Waterfall   Former position waterfall

RIVER'S COURSE

Hard rock

(A)

Section view

River

Waterfall

Hard rock

Gorge

(B)

Block diagram

A gorge will also form when a river maintains its course across a belt of country which is being uplifted. Only very powerful rivers are able to do this. The diagrams on page 51 show how this comes about. Notice that only parts of the region crossed by the river are uplifted and *not* the whole region. If the latter happened a gorge would develop along the entire length of the river.

The Indus, Brahmaputra and the headwaters of the Ganges have cut deep gorges in the Himalayas. The Indus gorge in Kashmir is 5180 metres (16940 feet) deep. The Columbia River has also cut a gorge across the Cascades in North America.

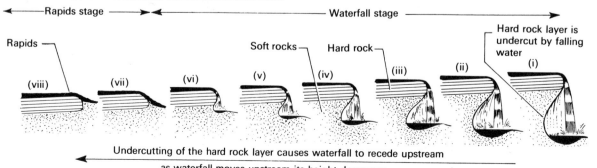

◄——— Rapids stage ———►◄——————— Waterfall stage ———————►

Rapids

(viii) (vii) (vi) (v) (iv) (iii) (ii) (i)

Soft rocks   Hard rock

Hard rock layer is undercut by falling water

◄—— Undercutting of the hard rock layer causes waterfall to recede upstream

as waterfall moves upstream its height decreases

The gorge below the Victoria Falls on the River Zambesi (Rhodesia). It is normally obscured by spray from the waterfall

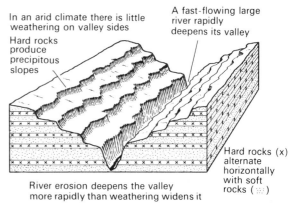

In an arid climate there is little weathering on valley sides

A fast-flowing large river rapidly deepens its valley

Hard rocks produce precipitous slopes

Hard rocks (x) alternate horizontally with soft rocks (⠿)

River erosion deepens the valley more rapidly than weathering widens it

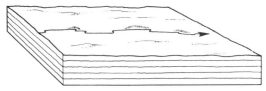

(i) Before formation of mountains

(ii) After formation of mountains

Vertical erosion by the river enables it to maintain its course across the rising mountains

Gorge   Gorge

Folded rocks

Earth movements have resulted in the formation of mountain ranges across the river's course

An impressive canyon in Utah's National Park (U.S.A.). The canyon is 792 metres (2600 feet) deep.

When a river flows across a plateau which is composed of horizontal and alternate layers of hard and soft rock, the valley it cuts will be deep and narrow. If the region is arid then there will be little weathering on the valley sides and the gorge will be very impressive. The Colorado River has cut a gorge 1·6 kilometres (1 mile) deep and 480 kilometres (300 miles) long into the Colorado Plateau (U.S.A.). Because of its size the gorge is called a *canyon*.

Canyons usually occur in dry regions where large rivers are actively eroding vertically and where weathering of the valley sides is at a minimum. This means that the river valleys are deepened more than they are widened. Canyons of great size develop when the land over which they flow is being uplifted but at a rate which enables rivers to maintain their courses across the area of uplift.

HORIZONTALLY BEDDED ROCKS

The Grand Canyon in U.S.A.

River action along the banks and bed of a river bend

The following two diagrams on page 52 show the erosional and depositional actions of a river as it flows round a meander. The water flows round the meander in a corkscrew manner and this causes erosion to take place on the concave bank and deposition on the convex bank.

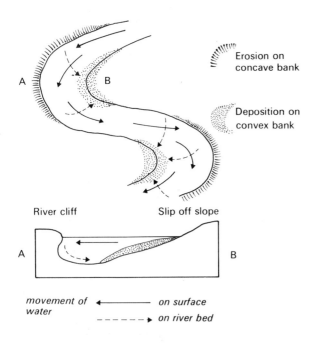

Erosion on concave bank

Deposition on convex bank

River cliff   Slip off slope

A   B

movement of   ⟵──── on surface
water   ──── ─▶ on river bed

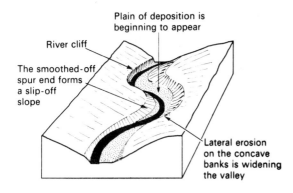

Plain of deposition is beginning to appear

River cliff

The smoothed-off spur end forms a slip-off slope

Lateral erosion on the concave banks is widening the valley

1 The young valley has well-developed interlocking spurs as shown in the diagram below. Lateral erosion on the concave bank has begun.

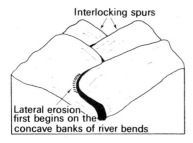

Interlocking spurs

Lateral erosion first begins on the concave banks of river bends

2 The valley is widened as the river meanders from side to side. Weathering lowers the valley sides. The river meanders migrate downstream and by doing so both widen and straighten the valley. Deposition takes place on the convex banks of meanders.

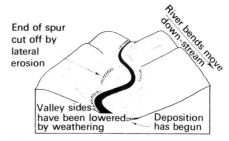

End of spur cut off by lateral erosion

River bends move down-stream

Valley sides have been lowered by weathering

Deposition has begun

3 The valley is now mature. The ends of spurs are cut right back and they stand up as *bluffs*.

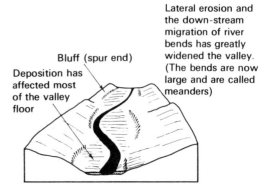

Lateral erosion and the down-stream migration of river bends has greatly widened the valley. (The bends are now large and are called meanders)

Bluff (spur end)

Deposition has affected most of the valley floor

4 The valley is now fully mature and it is approaching the stage of old age. Lateral erosion has developed a wide valley whose floor is almost completely covered with sediments. A flood plain is clearly being formed. The meander belt is as wide as the valley.

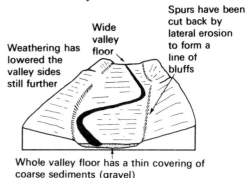

Wide valley floor

Spurs have been cut back by lateral erosion to form a line of bluffs

Weathering has lowered the valley sides still further

Whole valley floor has a thin covering of coarse sediments (gravel)

CHARACTERISTIC FEATURES OF A MATURE VALLEY

1 The valley has the shape of an open V in cross-section.
2 The gradient is more gentle than in a young valley.
3 River bends are pronounced. The concave banks stand up as river cliffs; the convex banks slope gently as slip-off slopes (smoothed ends of spurs).
4 Spurs are removed by lateral erosion. Their remains form a line of bluffs on each side of the valley floor.
5 The valley floor is wide and by the time the valley enters the stage of old age it is covered with a layer of sediments.

The building of the flood plain

1 During maturity the valley floor is widened by lateral erosion which is effected by meanders migrating downstream.
2 Active deposition begins to take place on the convex banks of meanders during maturity. Ultimately the whole valley floor is affected as meanders wander across it.

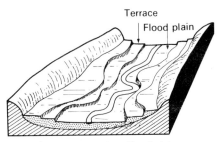

Meander terraces cut into flood plain.

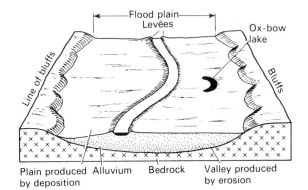

those on the other side. They are not paired as are terraces formed through *rejuvenation* (page 61).

Characteristic features of a flood plain

1 The river carries a heavy load some of which is deposited on its bed. This may produce mounds which divide the river into several channels. When this happens the river is said to be *braided*.
2 Meanders are pronounced and cut-offs develop and produce *ox-bow lakes*.

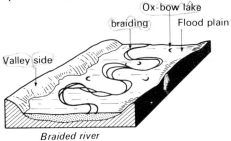

Braided river

3 After the stage of maturity is reached the river begins to overflow its banks and it deposits fine silts and muds on the valley floor. This is the final stage in the growth of a flood plain.

Note Meanders still migrate downstream and effect both lateral and vertical erosion. In doing so they may remove most of the original flood plain deposits though parts of these remain as terraces above the new flood plain which is developing. These meander terraces are of various heights and those on one side of the valley do not match in height

3 The river builds up its bed and banks with alluvium (the banks are called *levées*). The river thus flows between pronounced banks and above the level of the flood plain.

Ox-bow river entering Lake Athabasca in Canada

4 The river mouth sometimes becomes blocked with sediments and a delta forms.

The Development of an Ox-bow Lake

(i)

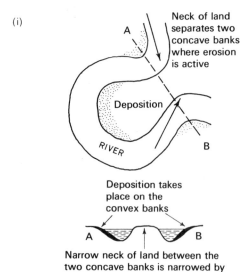

Neck of land separates two concave banks where erosion is active

Deposition

RIVER

A

B

Deposition takes place on the convex banks

Narrow neck of land between the two concave banks is narrowed by erosion

An acute meander where a narrow neck of land separates two concave banks which are being undercut.

(ii)

The neck is ultimately cut through. This may be accelerated by river flooding

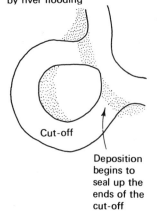

Cut-off

Deposition begins to seal up the ends of the cut-off

Erosion has broken through the neck of land. This often happens when the river is in flood. The meander has been cut off.

(iii)

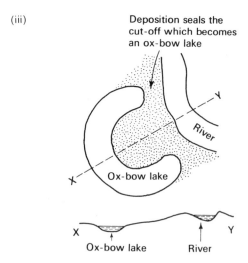

Deposition seals the cut-off which becomes an ox-bow lake

River

Ox-bow lake

X

Y

Ox-bow lake

River

Deposition takes place along the two ends of the cut-off and it is eventually sealed off to form an ox-bow lake.

Note how the area of deposition along the convex banks is increasing. After the formation of the ox-bow lake the river bed and banks are steadily raised by deposition and ultimately the river lies above the level of the ox-bow lake.

Mekong Meander

The photograph shows two pronounced meanders and an ox-bow lake (bottom left). The flood plain is clearly shown. Natural levées occur along both banks of the river. These are fairly conspicuous in the centre of the photograph.

The Formation of Levées and the Raising of the River Bed

1 Active deposition takes place along the banks of an old river when it is in flood. Each time this happens the banks get higher and they are called natural levees.

A meandering river in Sabah (East Malaysia) which has produced numerous ox-bow lakes.

2 When the river is not in flood deposition takes place on the river's bed. The bed is thus raised.

3 In time the river flows between levées and it is above the general level of the flood plain.
The Hwang-ho and Yangtze-kiang in China; the Mississippi* in the U.S.A. and the River Po in Italy all flow above the level of the flood plain in their lower courses.

* The bottom of this river is, however, below sea level.

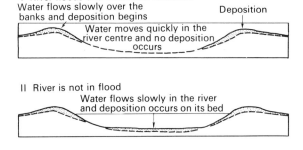

I River floods and overflows its banks
Water flows slowly over the banks and deposition begins
Deposition
Water moves quickly in the river centre and no deposition occurs

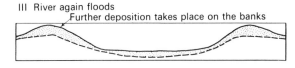

II River is not in flood
Water flows slowly in the river and deposition occurs on its bed

III River again floods
Further deposition takes place on the banks

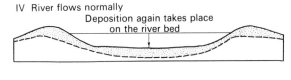

IV River flows normally
Deposition again takes place on the river bed

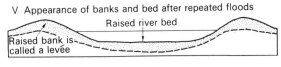

V Appearance of banks and bed after repeated floods
Raised river bed
Raised bank is called a levée

The Influence of Levels and Raised River Beds

When a river flows above the level of its flood plain, its tributaries find difficulty in joining it. They often parallel the river for many miles and in doing so they frequently meander themselves. They may cross depressions in the flood plain which then become swampy. Tributaries whose confluence with the main river is interfered with in this way are called *deferred junctions*.

Rivers which flow above the level of their flood plains are a constant menace. In times of severe floods they sometimes burst through the levées and disastrous floods spread out over the flood plains. The Hwang-ho, Yangtze-kiang and Mississippi all flow above the level of their flood plains in their lower valleys, and all periodically produce disastrous floods.

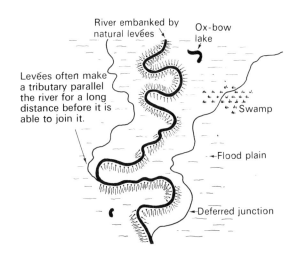

River embanked by natural levées
Ox-bow lake
Levées often make a tributary parallel the river for a long distance before it is able to join it.
Swamp
Flood plain
Deferred junction

Deltas

Most of the load carried by a river is ultimately dumped into the sea or lake into which it flows. The deposited load sometimes collects in the river mouth where it builds up into a low-lying swampy plain called a *delta*. As deposition goes on in the river mouth, the river is forced to divide into several channels each of which repeatedly divides. All these channels are called *distributaries*. Stretches of sea or lake become surrounded by deposited sediments and these are filled in with sediments when they may persist for some time as swamps. *Spits* and *bars* develop along the front of the delta. (Explanations on spits and bars are given later in the book).

There are Three Basic Types of Delta

1 *Arcuate*

This type is very common. It is composed of coarse sediments such as gravel and sand and is triangular in shape (see diagram of Nile Delta on page 56).

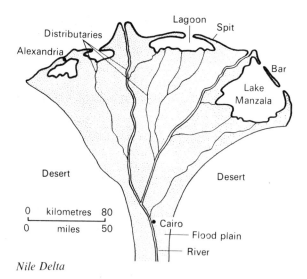

Nile Delta

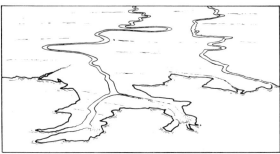

Vardar Delta

It always has a great number of distributaries. Rivers having this type of delta are: Nile, Ganges, Indus, Irrawaddy, Mekong, Hwang-ho and Niger.

2 Bird's Foot or Digitate.

This type is composed of very fine sediments called *silt*. The river channel divides into a few distributaries only and these maintain clearly defined channels across the delta. The Mississippi Delta is one of the best examples of this type. This is a diagram of the Vardar Delta. Two main distributaries can be seen. Both of these are flowing between levées which are another characteristic feature of this type of delta. This type occurs in seas which have few currents and tides to disturb the sediments.

3 Estuarine

An estuarine delta develops in the mouth of a submerged river. It takes the shape of the estuary. The deltas of the Elbe (Germany), Ob (U.S.S.R.) and the Vistula (Poland) are of this type.

Stages in the Development of a Delta

In stage 1 deposition divides the river mouth into several distributaries. Spits and bars arise and lagoons are formed. The levées of the river extend into the sea via the distributaries.

In stage 2 the lagoons begin to get filled in with sediments, and they become swampy. The delta

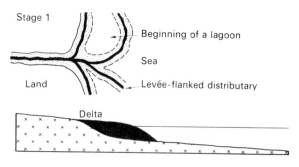

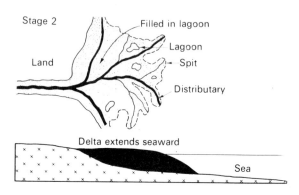

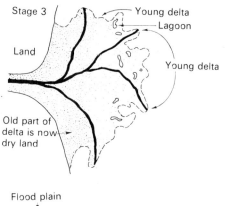

begins to assume a more solid appearance.

In stage 3 the old part of the delta becomes colonised by plants and its height is slowly raised as a result of this. Swamps gradually disappear and this part of the delta becomes dry land.

Note All three stages can often be seen in most deltas. As a delta grows larger and larger, the old parts merge imperceptibly with the flood plain, and they no longer have the appearance of a delta. Much of the North China Plain is the deltaic plain of the Hwang-ho; the plains of Iraq are the deltaic plains of the Tigris-Euphrates Rivers, and so on.

Conditions necessary for the Formation of a Delta
1 The river must have a large load, and this will happen if there is active erosion in the upper section of its valley.
2 The river's load must be deposited faster than it can be removed by the action of currents and tides.

Note
1 Deltas can, and do form on the shores of highly tidal seas, e.g. River Colorado (Gulf of California), and River Fraser (British Columbia).
2 Any river, irrespective of its stage of development, can build a delta. The Kander, whose valley is in the stage of youth, has built a delta in Lake Thun (Switzerland).

The Rate at which a Delta Grows
The formation of a delta results in an extension of the flood plain. Some deltas grow more rapidly than others. The Mississippi Delta is being extended seawards by 75 metres (245 feet) each year. The River Po (Italy) is extending its delta seawards by 12 metres (39 feet) per year. The advance of the Hwang-ho Delta has resulted in an island becoming joined to the mainland thus forming a peninsula (Shantung).

The delta of the Niger (West Africa)

Some Common Deltas from Asia and Africa

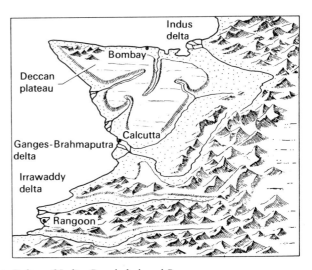

Deltas of India, Bangladesh and Burma

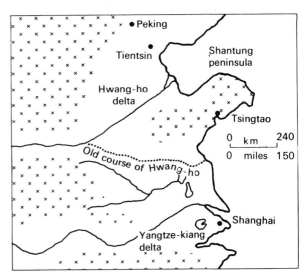

Hwang-ho Delta

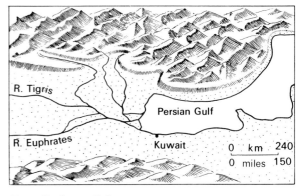

Tigris-Euphrates Delta

57

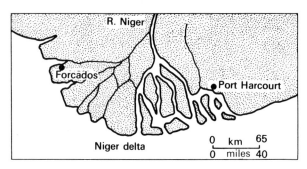

Niger Delta

The Value of Rivers and their Valleys to Man
Rivers
1 Some rivers, especially those in the stage of old age, form natural routeways which can be used for transport. The Yangtze-kiang and Mississippi and Rhine are particularly important.
2 Many rivers can be used for supplying irrigation water to agricultural regions, e.g. Nile, Tigris-Euphrates, Indus and Yangtze-kiang (irrigation will be discussed more fully in a later section).
3 For the development of hydro-electric power (H.E.P.). Young rivers which have waterfalls or which flow through gorges offer possibilities for H.E.P. development, e.g. of H.E.P. sites: *Kariba Dam* across the end of the gorge below the Victoria Falls on the River Zambesi; *Boulder Dam* near the Grand Canyon on the River Colorado; and *Owens Dam* near the Owens Falls on the River Nile where it leaves Lake Victoria. A dam is being built across the *Sanmen Gorge* on the Hwang-ho. Mature rivers can also be used for H.E.P. development, e.g. the dams at *Dnepropetrovsk* on the Dnieper, *Tsimlyanskaya* on the River Don, and at *Kuybyshev* on the River Volga all of which are in the U.S.S.R.
4 Some river mouths contain deep sheltered water and enable ports to be developed there, e.g. *Calcutta* on a branch of the Ganges, *Alexandria* on a distributary of the Nile, *Shanghai* in the delta of the Yangtze-kiang and *New Orleans* in the Mississippi Delta.

Valleys
1 Young valleys extend into highland regions by headward erosion. In doing so they often develop gaps by river capture. These gaps together with river gaps offer fairly easy passageways across the highlands. Roads and railways often take advantage of these.
2 Mature valleys by virtue of their wide floors offer gently sloping routeways to roads and railways across highland areas.
3 Mature valleys also offer good sites for settlements. Bluffs and river cliffs are frequently used as settlement sites.

4 The flood plains and deltas of old valleys contain fertile soils and provide Man with some of the best agricultural land. This is especially true of the subtropics and humid tropics where flood plains and deltas have for long been the home of large populations. It was on the riverine plains of the Nile, Tigris-Euphrates, Indus and Hwang-ho that the early civilisations grew up. All of these except the Hwang-ho have remarkably similar physical environments:
  (i) each is bordered by mountains, or deserts, or both
  (ii) each is open to the sea on one side
  (iii) each has a mild winter, a hot summer and an abundance of river flood water.

In Asia the flood plains and deltas form the most important physical landscapes, because most of the people live here. Wherever temperature and water conditions are suitable they are used for growing rice. The deltas of the Yangtze-kiang (China), Red River (N. Vietnam), Mekong (S. Vietnam), Irrawaddy (Burma) and the Indus (Pakistan) are especially important rice-growing regions. Outside Asia the Nile Delta, the Mississippi Delta and the Rhine Delta are also of great agricultural value.

Flood plains and deltas have their disadvantages. Sometimes their rivers cause serious flooding. This is particularly true when a river flows above the level of its flood plain and bursts its levées. Damage is done to crop land, settlements and communications. There may be considerable loss of life as there was in 1887 when the Hwang Ho burst its banks. Over 1 000 000 people lost their lives on that flood plain and delta in that year. The dangers of such serious flooding can be lessened by strengthening the levées and dredging the river to deepen its channel. The construction of dams across rivers also lessens the danger of flooding e.g. as in the Nile Valley, and if reafforestation is practised around the headwaters, less silt will be brought down.

DRAINAGE PATTERNS
All rivers are joined by smaller rivers or streams which are called *tributaries*. The area drained by a river and its tributaries is known as a *river basin* and its boundary is formed by the crest line of the surrounding highland. This boundary forms the main *watershed* of the basin.

A river system usually develops a pattern which is related to the general structure of its basin. Three distinct river patterns can be recognised. They are:
  (i) Dendritic
  (ii) Trellis
  (iii) Radial
A *dendritic* pattern develops in a region which is made of rocks which offer the same resistance to erosion and which has a uniform structure. The

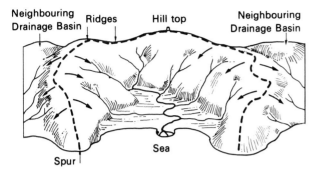

--- Watershed around the drainage basin
in the centre of the diagram

Diagram of a Watershed

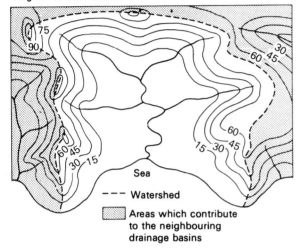

Sea

--- Watershed

Areas which contribute
to the neighbouring
drainage basins

at right angles to the general slope, down which
the principal river flows. The tributaries of the
main river extend their valleys by headward erosion
into the weak rocks which are turned into wide
valleys, called *vales*, and the hard rocks stand up
as *escarpments*.

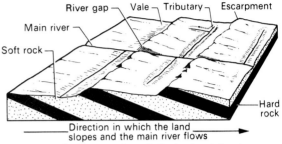

The main river cuts through the hard rock forming
the escarpment and a river gap is produced

The principal river which flows down the slope is
called a *consequent river* (C), and the tributaries
which cut out the vales and which do not flow
down the main slope are called *subsequent rivers*
(S).

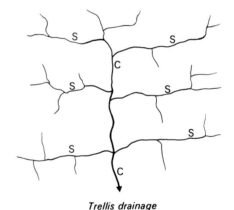

Trellis drainage

directions of the river and its tributaries are determined by slope. The land between main valleys and
between these and tributary valleys stands up as
ridges and spurs. The crests of these form the
watersheds. The river and its tributaries make a
pattern like the veins of a leaf.

Dendritic drainage

A *trellis* drainage pattern develops in a region which
is made up of alternate belts of hard and soft rocks
which all dip in the same direction and which lie

Dendritic Drainage

A *radial* drainage pattern develops on a dome or volcanic cone. The rivers flow outwards forming a pattern like the spokes of a wheel.

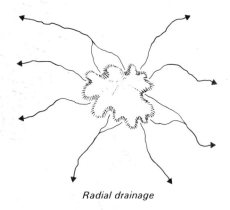

Radial drainage

River Capture

These diagrams show how river capture can take place. C_1 and C_2 are consequent rivers and S is a subsequent tributary of C_1. If S is a more powerful river than C_2 then it will lower its valley more rapidly than will C_2. This in turn will cause its tributary (S) to effect headward erosion and in time it will cut back into the valley of C_2 and effect river capture. When this happens the upper part of C_2 will flow into S and hence into C_1 thus making it an even more powerful river. The lower part of C_2 is now deprived of its headwaters and its volume decreases and it becomes too small for its valley. It is called a *misfit*. At the elbow of capture the valley now contains no river and it becomes a *wind gap*.

This process can be repeated until several parallel rivers have had their headwaters captured by which time C_1 will be the dominant river of the region.

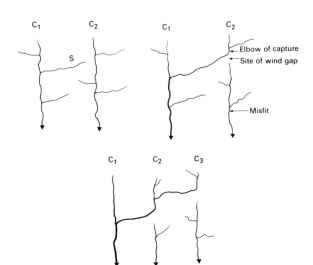

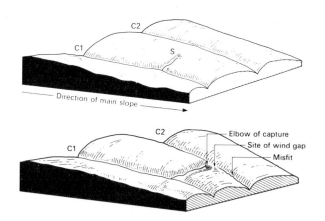

Direction of main slope

Elbow of capture
Site of wind gap
Misfit

Superimposed drainage patterns

Some rivers have developed a drainage pattern which is in no way related to the structure of the region in which it occurs. This happens when the drainage pattern was developed on a surface which overlay the present one. As the drainage pattern developed it cut its way into the underlying rock surface without regard for its structure and we say that it became superimposed on it. Gradually the whole of the original surface may be removed by erosion. In time the drainage pattern, especially the tributaries become affected by structure. The following diagrams show the development of a superimposed drainage pattern.

1 *Original folded surface*

Strong rock

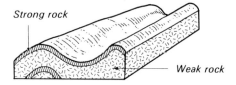

Weak rock

2 *Region is reduced to a peneplain*

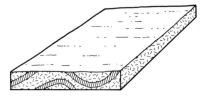

3 *Subsidence results in region being buried by newer rocks but subsequent uplift sees the formation of a drainage pattern. The main rivers are draining at right angles to the axis of the original structure.*

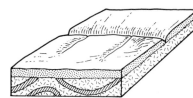

4 Tributaries to the main river develop wide valleys in the weaker rocks. As the main river erodes vertically it cuts across the ridges of strong rock and forms gorges. The strong rock forms ridges because the weak rocks are worn away and not because the region has been uplifted.

Rejuvenation

When the base level (page 47) of a river is lowered. the river's power to erode is increased. Most of the erosion is downwards into the valley floor and as a result new types of features are produced. A river which is given this extra power is, in effect, given new life, that is, it is *rejuvenated*.

The lowering of a river's base level may be caused by an uplift of the land or by a fall in the level of the sea. The latter happens when water is withdrawn from the seas to be locked up in ice masses when the climate of some regions gets colder, for example, as in ice ages.

A river in any stage of development from youth to old age may be rejuvenated which means that the valley eroded by a rejuvenated river, which will be a young valley, may occur in an old landscape. For example, if a river crossing a peneplain is rejuvenated, its valley will have the typical V-shape of a young valley.

Some features produced by rejuvenated rivers

1 If a river on a flood plain is rejuvenated, the downcutting effected by the river will produce terraces. These will be paired, that is the heights of the terraces on one side of the river will correspond with the heights of the terraces on the other side. These terraces are different from meander terraces (page 53). It sometimes happens that the point where the river crosses from the original flood plain to the new flood plain is visible and is marked by rapids or a waterfall. This point is called the *knickpoint*.

Knickpoint

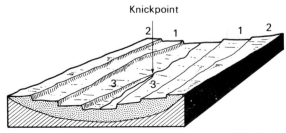

Paired rejuvenation terraces (11, 22, 33) and knickpoint.

2 If a river is able to maintain its course over a region of uplift then the river will rapidly deepen its valley. If the rocks of the region are resistant and if the climate is dry, then a very impressive gorge can develop. The Grand Canyon is a good example (page 51).

EXERCISES

1 Carefully distinguish between the following:
    (i) weathering by temperature changes and weathering by frost action
    (ii) a scree and an exfoliation dome.

2 What are the main differences between each of the following:
    (i) a porous rock and a pervious rock
    (ii) a permeable rock and an impermeable rock
Name *one* specific example of each type of rock and say which condition, if it occurs in surface rocks, is likely to promote flooding.

3 Briefly describe the manner in which a land surface can be changed by: (i) temperature changes, (ii) rain, (iii) running water. Illustrate your answer with annotated diagrams and give specific examples.

4 Draw fully labelled diagrams to show the differences between: (i) a spring, and (ii) a well. Write brief notes on each and state in what type of region either, or both, may be found.

5 By means of well-labelled diagrams, and supporting text, show how (i) an oasis, and (ii) an artesian basin, may occur.

6 Explain, by well-labelled diagrams, the main differences between the following pairs of features:
    (i) waterfall and rapid
    (ii) distributary and tributary
    (iii) radial and dendritic drainage pattern.

7 Describe the characteristic features which commonly occur in a river valley during the stages of youth, maturity and old age. Illustrate the more important features by means of well-labelled diagrams.

8 Name *two* features produced by river erosion and *two* features produced by river deposition. For each feature state:
    (i) the type of valley in which it occurs
    (ii) where it occurs in that valley.
For *one* of the erosional and *one* of the depositional features which you have named, draw a well-labelled diagram to show its main characteristics and say how the feature may have formed.

9 Choose any *two* pairs of the following features and for *each*:
(a) say how the features develop, (b) name *one* valley where the features occur, (c) draw well-labelled diagrams of the features.
    (i) levée and flood plain
    (ii) matched river terraces and unmatched river terraces
    (iii) ox-bow lake and cut-off

Objective Exercises

1 Which of these statements best describes the meaning of weathering?
   A the burrowing action of animals and the growth of plant roots
   B the freezing of water in cracks in rocks
   C the break-up of rocks exposed at the surface
   D the alternate heating and cooling of rocks
   E the dissolving out of mineral particles by rain water

   A B C D E
   ☐ ☐ ☐ ☐ ☐

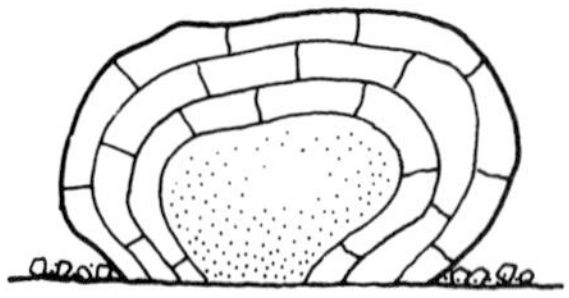

2 The diagram above shows a common type of weathering in a hot desert. This type is known as
   A attrition
   B deflation
   C extrusion
   D solution
   E exfoliation

   A B C D E
   ☐ ☐ ☐ ☐ ☐

3 A well will always contain water if
   A it is sunk at the bottom of a hill
   B the bottom of the well is far below the water-table
   C it is sunk into sedimentary rocks
   D it is located in a rainy region
   E it is on a spring line

   A B C D E
   ☐ ☐ ☐ ☐ ☐

4 All the following features are produced by denudation. Which one is produced by rain action?
   A gully
   B gorge
   C earth pillar
   D exfoliation dome
   E cliff

   A B C D E
   ☐ ☐ ☐ ☐ ☐

5 The formation of a river delta involves the following processes. In what order do they take place?
   A transport, corrasion, deposition
   B corrasion, transport, deposition
   C deposition, corrasion, transport
   D corrasion, deposition, transport
   E deposition, transport, corrasion

   A B C D E
   ☐ ☐ ☐ ☐ ☐

6 The erosive power of a river depends most upon its
   A width and depth
   B speed and depth
   C gradient and width
   D speed and width
   E speed and volume

   A B C D E
   ☐ ☐ ☐ ☐ ☐

7 Deposition by a river increases when
   A the volume of water in the river increases
   B the river enters a gorge
   C the river flows more quickly
   D the river enters a lake
   E the river approaches a waterfall

   A B C D E
   ☐ ☐ ☐ ☐ ☐

8 The load of a river comes mainly from
   A the banks which are undercut by the river on a meander
   B the valley sides down which rock particles are moved by soil creep
   C the bluffs which are undercut by the river in its upper course
   D the river's bed which has been abraded by the action of 'pot-holing'

   A B C D
   ☐ ☐ ☐ ☐

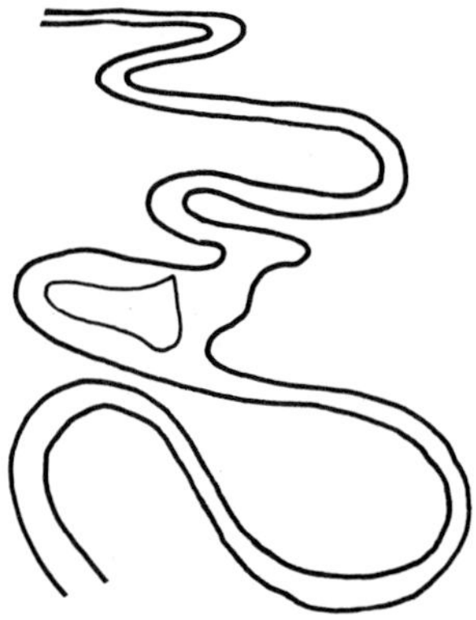

9 The diagram above is a drawing of a meandering river. Using this diagram suggest an area where an ox-bow lake is likely to be formed.
   A where the river current slows down
   B where deposition is taking place along a meander
   C where the meander neck is cut off
   D at a meander core

   A B C D
   ☐ ☐ ☐ ☐

10 Which of these features clearly indicates the presence of a flood plain?

A seasonal floods
B ox-bow lakes
C meanders
D distributaries

A B C D
□ □ □ □

11 Most rivers flow slowly near to sea level and in consequence their main action is depositional. A river in this stage would **not** show signs of

A a wide flat-floored valley
B deposits of large boulders
C river levees
D tributaries with deferred junctions

A B C D
□ □ □ □

12 Which one of the following combinations of natural conditions would best promote the formation of a delta at the mouth of a heavily loaded river (assuming there is no subsidence)?

A a saline sea with strong currents
B a fresh-water sea with negligible currents
C a fresh-water sea with strong currents
D a saline sea with negligible currents

A B C D
□ □ □ □

13 A drainage system which is in no way related to the structure of the region where it occurs is called a

A radial drainage pattern
B trellis drainage pattern
C superimposed drainage pattern
D dendritic drainage pattern

A B C D
□ □ □ □

14 All the following features **except** one may suggest that a drainage system has been rejuvenated

A knick-points
B paired river terraces
C flat-floored valleys
D incised meanders

A B C D
□ □ □ □

15 "The wearing away of the sides and bottom of a river's channel by the load carried by a river" is called

A corrosion
B attrition
C corrasion
D transportation

A B C D
□ □ □ □

16 Which one of the following statements best explains why the lower course of a river is sometimes choked with sediments

A The valley of a river is widest in its lower course.
B The velocity of a river in its lower course is low.
C A delta sometimes develops in a river's lower course.
D Waterfalls rarely occur in the lower course of a river.

A B C D
□ □ □ □

17 A region which has a trellis drainage pattern usually has the following features

A swallow holes and dry valleys
B wind gaps and vales
C erratics and moraines
D clints and grikes

A B C D
□ □ □ □

18 Which one of the following best explains the formation of canyons?

A a large volume of water
B greater vertical erosion by a glacier of a main valley than of the tributary valleys
C pronounced Earth movements
D vertical erosion in the valley of a river crossing an arid plateau

A B C D
□ □ □ □

19 Chemical weathering takes place most effectively when it is

A warm and wet
B cold and dry
C cold and wet
D hot and wet
E dry all the time

A B C D E
□ □ □ □ □

20 Under which one of the following conditions is physical weathering most effective?

A hot and dry all the time
B hot and wet all the time
C hot season alternating with dry season
D large diurnal temperature range
E even temperatures all the time

A B C D E
□ □ □ □ □

21 Which one of the following features is the product of river deposition?

A scree
B earth pillar
C landslide
D levée

A B C D
□ □ □ □

5   Limestone and Chalk Features

Limestone Landscape

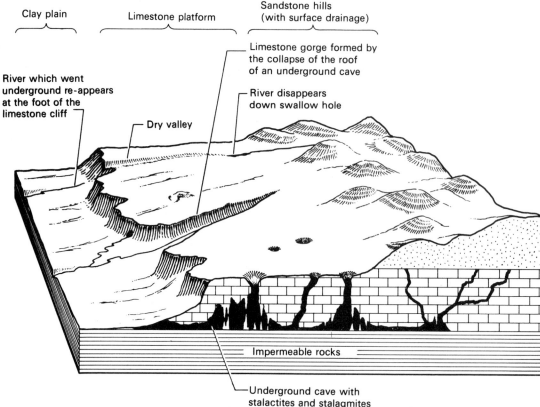

DRAINAGE FEATURES IN A LIMESTONE REGION

The nature of limestone
Limestone consists chiefly of calcium carbonate which is insoluble. The carbon dioxide, which rain water absorbs from the air, turns the insoluble carbonate into soluble bicarbonate. This is the reason why rain water and rivers are able to remove limestone in solution.

$$CaCO_3 + CO_2 + H_2O = Ca(HCO_3)_2$$
$$\text{carbonate} \qquad\qquad \text{bicarbonate}$$

Limestone is a well-jointed rock and its joints and bedding planes soon become opened up by rain and water, and in time the surface consists of broken and rugged rocks.

Limestone landscape
One of the most noticeable features of a limestone landscape is the almost complete absence of surface drainage. The permeability of limestone permits rain to soak into it very easily. Joints rapidly become

excavated and deepened, with the result that the surface becomes criss-crossed with wide irregular gullies, known as *grikes*. The intervening blocks of limestone surface are called *clints*.

Rivers rising in a non-limestone region sometimes flow into a limestone region. When this happens the rivers disappear into vertical holes in the surface and continue to flow as *underground rivers* inside the limestone. The vertical holes, called *swallow holes* or *sink holes*, are formed by rivers and they are usually widened vertical joints. Gaping Ghyll in Yorkshire, England, is a particularly good example. Swallow holes may join together to give a very large opening, called a *doline*. Likewise, dolines may join up to give even larger openings. These are called *uvala*.

Rivers which flow inside limestone develop underground caves and caverns as they flow along joints and bedding planes. Some caves are of great size, e.g. Carlbad Cave (New Mexico – U.S.A.) is 1200 metres (3950 feet) long, 183 metres (600 feet) wide and 90 metres (295 feet) high. Batu Caves, near

Gaping Gill near Ingleborough, Yorkshire, England.

Kuala Lumpur and the caves near Ipoh are further examples. *Stalagmites* and *stalactites* develop in these caves and sometimes they join together to form natural columns or pillars.

Sometimes the roof of an underground cave or cavern collapses and a gorge, with almost vertical sides, then develops. Cheddar Gorge (the U.K.) was formed in this way.

Water containing calcium bicarbonate drips from the cave roof. When the water evaporates it leaves behind calcium carbonate

Interior view of a limestone cave

Rivers which disappear underground when entering a limestone region reappear on the surface again where the junction of the limestone and the underlying impermeable rocks meet the surface. Dry, gorge-like valleys often mark the former courses of such rivers and these occur between the point of disappearance and the point of emergence (diagram, page 64). The former course over the limestone in European regions was probably made possible by the frozen subsoils in Glacial Times. Dry waterfalls also occur in these valleys, especially where the rivers once crossed limestone escarpments.

The surface of a limestone region is not only broken, it is also stony. Any soil which may occur is usually

Interior of Cango Caves (South Africa)

in small shallow patches which support only a few shrubs, grasses and in some regions sweet-smelling herbs. Larger plants, such as trees, only occur in the bottom of large valleys which have been excavated down to the rocks underlying the limestone. Although the limited plant life in limestone regions varies from region to region, it being dependent upon the nature of the climate, the general appearance of all limestone regions is very much the same. The limestone region around Ipoh, in Perak (Peninsular Malaysia), is well-covered with vegetation because of the fairly deep soils which have formed under humid tropical weathering.

Limestone landscapes are called *karst* landscapes and good examples occur in north-west Yugoslavia, the Pennines of the U.K., the Yucatan Peninsula of Mexico, the Kentucky region of the U.S.A. and parts of Perak and Perlis in Peninsular Malaysia.

Value of Karst regions to Man

Because of their barren nature karst regions contain few settlements. The dryness of the surface and the limited amounts of poor soils prevent the growth of a continuous plant cover. In some regions there is sufficient grass to support sheep or goats and animal grazing takes place. Occasionally areas of good soils do occur. These are usually confined to basins which have been formed by the collapse of roofs of underground caverns. In Yugoslavia and other parts of the Mediterranean region, these soils are usually red and are called *terra rossa*. They are valuable for farming.

Limestone is quarried as a building stone and for making cement, and usually there are stone and cement works near to limestone regions, e.g. near to Ipoh in Peninsular Malaysia.

Features of a chalk landscape

Chalk, like limestone, is made of calcium carbonate but it is much softer than limestone. Its surface is not marked by outcrops of hard rock. Instead it is usually gently undulating with rounded hills, called *downs* in England, and wide open valleys, which are usually without rivers. Chalk is a porous rock and rain falling on its surface rapidly soaks into the ground. There is, therefore, very little surface run-off, that is, there are very few streams. Because the valleys are without streams, they are called *dry valleys* or *coombs*.

Good examples of chalk landscapes occur in England in the Chiltern Hills and the Downs, and in these regions dry valleys are very common. These valleys were obviously formed when the water-table was higher than it is at present. Possibly, towards the end of the last glacial period, vast quantities of melt water from the retreating ice sheets were able to flow as rivers across these chalk regions, because

the subsoils were frozen, thus presenting an impermeable zone.

EXERCISES

1 Briefly distinguish between the following:
    (i) a dry valley and an underground river
    (ii) a limestone gorge and a swallow hole.
    (iii) a clint and a grike
    Name *one* region where these types of features may be seen.

2 Carefully explain why (i) some underground rivers produce varied underground scenery, (ii) most limestone areas have little agriculture and few people, and (iii) there is almost no surface drainage in a limestone region. Illustrate your answer with relevant diagrams.

Objective Exercises

1 Which one of the following is characteristic of a limestone region?
A dry valleys
B meandering streams
C deep soils
D good cover of natural vegetation
E salt marshes

    A B C D E
    ☐ ☐ ☐ ☐ ☐

2 All of the following features are likely to occur in a limestone region **except**
A underground rivers
B ox-bow lakes
C dry valleys
D clints
E caves

    A B C D E
    ☐ ☐ ☐ ☐ ☐

3 A karst landscape is most likely to develop in a region whose rocks are
A porous
B impermeable
C chiefly made of calcium carbonate
D well jointed

    A B C D
    ☐ ☐ ☐ ☐

4 A doline is
A a large crack produced by erosion on the surface of a limestone plateau
B a large swallow hole produced by 'solution'
C an underground cave
D the opening through which an underground river re-emerges at the surface.

    A B C D
    ☐ ☐ ☐ ☐

6 Features Produced by Wind

WIND ACTION AND THE FEATURES IT PRODUCES

Wind action is very powerful in arid and semi-arid regions where rock waste is produced by weathering and is easily picked up by the wind. In humid regions rock particles are bound together by water droplets and plant roots, and hence there is little wind erosion.

Wind erosion consists of *abrasion* which breaks up rocks and produces rock pedestals, zeugens, yardangs and inselbergs, and *deflation* which blows away rock waste and thus lowers the desert surface producing depressions, some of which are very extensive.

Wind transport causes fine particles of rock waste, called desert dust, to be carried great distances. Coarser particles called sand are bounced over the surface for short distances.

Wind deposition gives rise to dunes which are made of sand and loess which is made of desert dust.

Types of tropical and temperate desert surface

Most of the world's deserts are located in latitudinal belts of 15° to 30° north and south of the equator.

(a) West Coast Deserts

These mainly occur in the trade wind belt on the western sides of continents where the winds are off-shore. On-shore local winds do blow across these coasts but they rarely, if ever, bring rain because they have to cross cool currents which parallel the coasts in these latitudes. The cool currents cause condensation to take place in the on-shore winds which produces mist, fog or light rains. By the time they reach the coasts the winds are dry.

(b) Continental Deserts

These occur in the interior of continents where the winds have already travelled a considerable distance across the land and in doing so, have lost much of their moisture as happens when winds blow over a dry land or over high mountains. The day temperatures of these continental areas are very high because they are so far from the moderating influence of the sea. The night temperatures are low because the absence of clouds causes these areas to lose their heat rapidly. Deserts of this type occur mainly in west coast tropical latitudes but they do extend into the temperate zones. Sometimes these deserts occur in intermontane plateau regions. The deserts of Arizona and Nevada in the Rockies are of this type and they are sometimes known as Mountain Deserts. A desert often occurs because of a combination of these factors e.g. the Sahara and the Australian Desert are partly of the west coast type and partly of the continental type.

Although the dominant agent of erosion in deserts is wind, rain does occur which results in both water erosion and water deposition. The following types of desert surface can be recognised.

Sandy Desert – called *erg* in the Sahara and *koum* in Turkestan. This is an undulating plain of sand produced by wind deposition.

Stony Desert – called *reg* in Algeria and *serir* in Libya and Egypt. The surface is covered with boulders and angular pebbles and gravels which have been produced by diurnal temperature changes.

Rocky Desert – called *hamada* in the Sahara. The bare rock surface is formed by deflation which removes all the small loose rock particles. A part of the Sahara Desert in Libya has extensive areas of this type of desert.

Badlands – This is quite a different type of desert to the three just mentioned in that it develops in semi-desert regions mainly as a result of water erosion produced by violent rain storms. The land is broken by extensive gullies and ravines which are separated by steep-sided ridges. Excellent examples of this type of desert occur in the region extending from Alberta to Arizona (North America).

Features produced by Wind Erosion

Wind abrasion attacks rock masses and sculptures them into fantastic shapes. Some of these, because of their shape, are called rock pedestals.

I Rock Pedestals

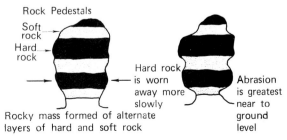

Rock Pedestals
Soft rock
Hard rock
Hard rock is worn away more slowly
Abrasion is greatest near to ground level
Rocky mass formed of alternate layers of hard and soft rock

Rock Pedestals in the Lut Desert in Iran

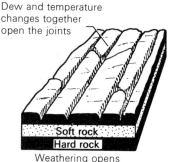

Dew and temperature changes together open the joints

Soft rock
Hard rock

Weathering opens up the joints

II Zeugens

Wind abrasion turns a desert area which has a surface layer of hard rock underlain by a layer of soft rock into a 'ridge and furrow' landscape. The ridges are called zeugens. A zeugen may be as high as 30 metres (100 feet). Ultimately they are undercut and gradually worn away.

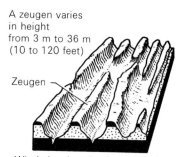

Wind abrasion develops furrows in the soft rocks

Hard rock forms block-like ridges called zeugens

Wind abrasion continues the work of weathering

III Yardangs

Bands of hard and soft rocks which lie parallel to the prevailing winds in a desert region are turned into another type of 'ridge and furrow' landscape by wind and abrasion. The belts of hard rock stand up as rocky ribs up to 15 metres (50 feet) in height and they are of fantastic shapes. They are called yardangs. Yardangs are very common in the central Asian deserts and in the Atacama Desert.

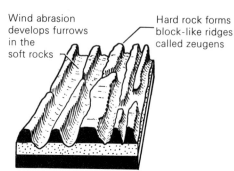

A zeugen varies in height from 3 m to 36 m (10 to 120 feet)

Zeugen

Wind abrasion slowly lowers the zeugens and widens the furrows

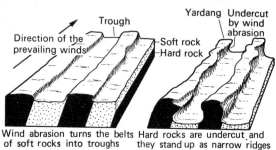

Direction of the prevailing winds

Trough

Yardang Undercut by wind abrasion

Soft rock
Hard rock

Wind abrasion turns the belts of soft rocks into troughs

Hard rocks are undercut and they stand up as narrow ridges called yardangs

IV Depressions

Some depressions produced by wind deflation reach down to water-bearing rocks. A swamp or an oasis then develops.

The Qattara Depression is 122 metres (400 feet) below sea level. It has salt marshes and the sand excavated from it forms a zone of dunes on the leeside.

Fault-produced depression

The formation of a depression may first be caused by faulting. The soft rocks thus exposed are excavated by wind action.

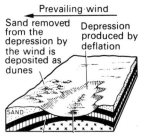

Prevailing wind

Sand removed from the depression by the wind is deposited as dunes

Depression produced by deflation

SAND

Aquifer Water seeps out of aquifer and forms swamp or an oasis

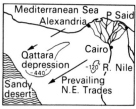

Mediterranean Sea
Alexandria
P Said
Cairo
Qattara depression
-440
R. Nile
Sandy desert
Prevailing N.E. Trades

The Qattara Depression is 122 m (400 ft) below sea level. It has salt marshes and the sand excavated from it forms a zone of dunes on the leeside.

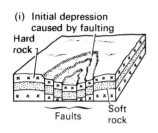

(i) Initial depression caused by faulting

Hard rock

Faults   Soft rock

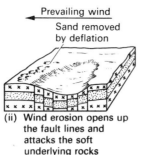

Prevailing wind
Sand removed by deflation

(ii) Wind erosion opens up the fault lines and attacks the soft underlying rocks

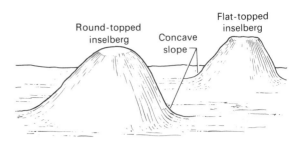

Round-topped inselberg   Flat-topped inselberg   Concave slope

Inselberg

In some desert regions erosion has removed all of the original surface except for isolated pieces which stand up as round-topped masses, called inselbergs. Some of them are probably the remains of plateau edges which have been cut back by weathering after which the weathered rock waste has been removed by sheet wash. Others may be the result of wind erosion or the combined action of wind and water erosion.

Inselbergs are common in the Kalahari Desert, parts of Algeria, north-west Nigeria and Western Australia.

Features produced by Wind Deposition

Strong winds occasionally blow across desert surfaces and when they occur they carry vast amounts of desert dust (very fine sand particles). This movement produces dust storms and in time it results in the transport of enormous quantities of fine material from one part to another part of a desert, or from a desert to a neighbouring region. Slight movements of air, called *wind eddies*, bounce grains of sand forward, which, when deposited, may form gentle ripples or sandy ridges, called *dunes*.

Sand dunes

There are two types of dune, *barchan* (barkhan) and *seif*. A barchan is crescent-shaped, lies at right angles to the prevailing wind and is much more common than a seif which is long and usually straight, and which is parallel to the prevailing wind. A barchan usually develops from the accumulation of sand caused by a small obstruction such as a rock or a bit of vegetation. As the mound of sand grows bigger its two edges are slowly carried forward down-wind and the typical crescent shape develops. The windward face of the dune is gently-sloping but the leeward side is steep and is slightly concave. This is caused by wind eddies which are set up by the prevailing wind (see diagram). A barchan moves slowly forward as grains of sand are carried up the windward face and slip down the leeward side. Barchans range in height from a few metres to over thirty metres and they may occur singly or in groups. There are good examples

Ayers Rock, Northern Territory, Australia. This inselberg is 1600 metres (1750 yards) long, 985 metres (1077 yards) wide and 340 metres (372 yards) high.

in the deserts of the Sahara and Turkestan.

A *seif dune* forms when a cross wind develops to the prevailing wind and the corridors between the dunes are swept clear of sand by this wind. Eddies blow up against the sides of the dunes and it is these which drop the sand grains and thus build up the dunes. The dunes are lengthened by the prevailing wind. Seif dunes are often several hundred metres high and many kilometres long. Some even reach a length of 160 kilometres (100 miles). Good examples of seif dunes occur in the Thar Desert, the desert of Western Australia and south of the Qattara Depression.

Loess Deposits

The wind blows fine particles out of the deserts each year. Some are blown into the sea, the rest are deposited on the land where they accumulate to form loess. Loess is friable and easily eroded

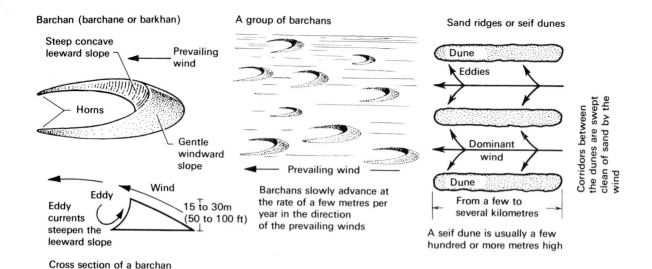

Barchan (barchane or barkhan)

Steep concave leeward slope

Prevailing wind

Horns

Gentle windward slope

Eddy currents steepen the leeward slope

Eddy

Wind

15 to 30m (50 to 100 ft)

Cross section of a barchan

A group of barchans

Prevailing wind

Barchans slowly advance at the rate of a few metres per year in the direction of the prevailing winds

Sand ridges or seif dunes

Dune

Eddies

Dominant wind

Dune

From a few to several kilometres

Corridors between the dunes are swept clean of sand by the wind

A seif dune is usually a few hundred or more metres high

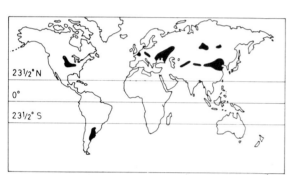

Loess regions of the world

23½° N

0°

23½° S

In northern China the loess has been intensely eroded by rivers to give a 'badland' landscape. The photograph below shows deep gorge-like valleys which have been cut into the loess. The centre of the photograph shows a loess plain which is under cultivation.

Many of the loess hills are terraced for crops. Most of the people in this loess region live in houses which have been carved out of the loess cliffs.

by rivers. There are extensive loess deposits in northern China. These are formed of desert dust blown out of the Gobi Desert to the west. The loess deposits of central Europe were probably deposited in the last Ice Age when out-blowing winds carried fine glacial dust from the ice-sheets of northern Europe. The loess deposits of the Pampas have been derived from the deserts to the west.

Loess cave dwellings of Shansi Province in Northern China. These cave dwellings are warm in winter and cool in summer; however just slight earthquakes can cause serious landslides and the consequent loss of many lives.

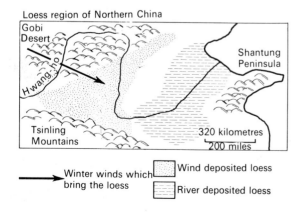

Loess region of Northern China

Gobi Desert

Hwang-ho

Shantung Peninsula

Tsinling Mountains

320 kilometres
200 miles

→ Winter winds which bring the loess

▦ Wind deposited loess

▤ River deposited loess

WATER ACTION IN DESERTS

A desert region may receive no rain for several years and then a sudden downpour of from 100 to 250 millimetres (3·9 to 9·75 in) occurs. These rare but heavy rain storms give birth to rushing torrents on steep slopes and sheet flood water on gentle slopes. The run-off on steep slopes is usually via *rills* (shallow grooves) which lead into gullies which in turn connect with steep-sided, deep and often flat-floored valleys, called *wadis* or *chebka* (in Algeria). During sudden rain storms, flood waters rush down wadis as *flash floods*. These are short-lived and the large quantities of sediment which they carry are deposited in bulk giving rise to alluvial fans and delta-like deposits at the foot of steep slopes, e.g. where a tributary joins a wadi, and where the flood waters dry up, e.g. at the exit of a wadi.

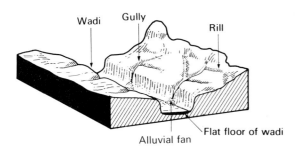

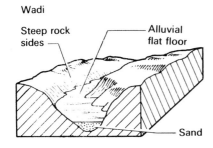

In intermontane desert basins, e.g. Tarim Basin, intermittent rivers drain into the centre of the basin. The alluvial fans which build up around the edge of such a basin may eventually join together to form a continuous depositional feature, sloping gently to the centre of the basin. This feature is called a *bahada* or *bajada*. Sometimes the centre of a basin is occupied by temporary salt lakes, caused by high evaporation, e.g. the shotts of North Africa and playas of North America. In some cases the lakes

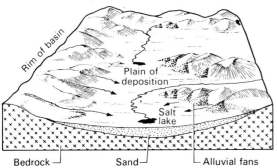

Crescent Dunes in North Africa

dry up and only the salt beds of these remain. These are called *salinas*.

Intermontane Desert Basin

As the edges of desert and semi-desert highlands get pushed back by erosion and weathering, a gently sloping platform develops. This is called a *pediment*. The slope of the land changes abruptly where a pediment joins the highland mass.

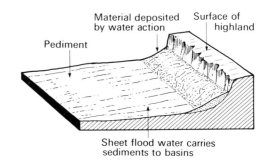

Wadis and Rocky Ridges in Jordan

Mount Sinai in Egypt

Reg of Sinai

Deserts and rivers

Most deserts are regions of inland drainage, i.e. their rivers and streams never reach the sea. Very few rivers persist throughout the year in deserts but there are some significant exceptions. The Nile, in North Africa, the Tigris-Euphrates, in Southwest Asia, and the Colorado, in the U.S.A., are three of the best examples of rivers which cross desert areas and which are permanent rivers. This is because these rivers rise in regions of rain which falls throughout the year and which is sufficient to sustain a permanent flow of water across the desert areas.

The Value of Wind-blown Deposits to Man

1 Loess deposits are unusually fertile. In northern China the loess region has been cultivated for 4000 years. In the Ukraine (north of the Black Sea), in the Pampas of Argentina, and in the High

This map shows those deserts where wind action is important.

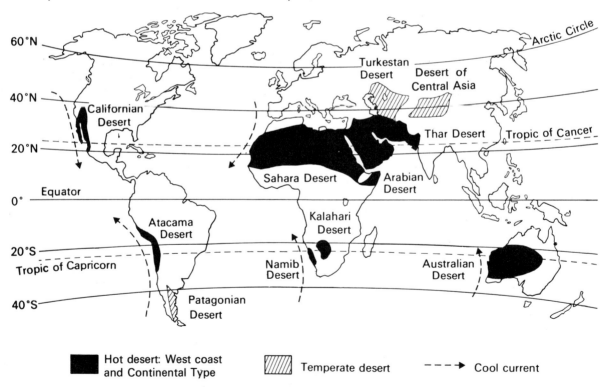

Plains of the U.S.A., the loess soils are very fertile and are used for grain growing.

2 Loess deposits of both China and Europe are used as a means for building dwellings. These are cut into the deposits. Their chief advantages are the ease with which they are built and their warmness in winter and their coolness in the summer. Their main disadvantage is their instability (they often collapse during even mild earthquakes).

EXERCISES

1 For each of any *three* of the following features:
   (a) rock pedestal
   (b) yardang
   (c) zeugen
   (d) inselberg
   (e) deflation depression
      (i) draw a large diagram to show the main characteristics of the feature, and name these
      (ii) carefully explain how the feature was produced
      (iii) name *one* region where an example of the feature may be seen.

2 Name *two* types of depositional feature which develop under arid conditions and for *one* of these briefly explain how it forms.

Objective Exercises

1 Wind can effect erosion through the process of
   A abrasion
   B exfoliation
   C extrusion
   D solution
   E attrition

   A B C D E
   □ □ □ □ □

2 Which one of the following features is **not** formed by wind erosion
   A yardang
   B rock pedestal
   C zeugen
   D barchan
   E hamada

   A B C D E
   □ □ □ □ □

3 Which of the following features has been produced by wind deflation?
   A Lake Toba (Sumatra)
   B Lake Chad
   C Lake Baikal
   D Lake Victoria
   E Qattara Depression

   A B C D E
   □ □ □ □ □

4 Loess consists of fine rock particles which
   A are often deposited on the flood plains of large river valleys
   B frequently occur in regions adjacent to deserts which lie under winds blowing from the deserts
   C result from the weathering of limestone rocks
   D are deposited by waves along arid coastlines
   E are formed from the weathering of volcanic rocks

   A B C D E
   □ □ □ □ □

5 Which one of the following landforms is **not** the product of denudation in a hot arid region?
   A seif
   B reg
   C grike
   D erg

   A B C D
   □ □ □ □

6 Although hot deserts have a very low annual rainfall, occasional heavy rain storms do occur and these sometimes produce steep-sided, flat-floored valleys called
   A gorges
   B wadis
   C canyons
   D dry valleys

   A B C D
   □ □ □ □

7 Two types of plains frequently occur between the borders of desert basins and the surrounding highlands. They are called bahadas (bajadas) and pediments. Bahadas are
   A alluvial fans which have joined together to form a sloping plain around the desert basin
   B located at the entrance to a wadi
   C sloping rock platforms produced by wind erosion
   D the steep slopes of the surrounding highlands

   A B C D
   □ □ □ □

8 A pediment is
   A a rocky surface
   B made of loose gravel
   C the remnant of receding mountain slopes
   D composed of deposits of sediments

   A B C D
   □ □ □ □

9 Which of the following rivers crosses an extensive desert before it reaches the sea?
   A Mississippi
   B Amazon
   C Hwang-ho
   D Colorado

   A B C D
   □ □ □ □

7   Features Produced by Waves

SEA ACTION AND THE FEATURES IT PRODUCES

Coasts are forever changing—some are retreating under wave erosion and others are advancing under wave deposition. There are many different types of coasts; some are steep, some are gentle, some are sandy, some are rocky. The character of a coast results from two or more of the following factors:

1 Wave action
2 Nature of the rocks forming the coast
3 Slope of the coast
4 Changes in the level of sea or land
5 Volcanic activity
6 Coral formations
7 The effects of glaciers
8 The action of man.

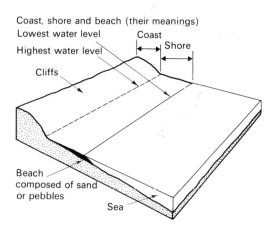

Coast, shore and beach (their meanings)
Lowest water level
Highest water level
Coast
Shore
Cliffs
Beach composed of sand or pebbles
Sea

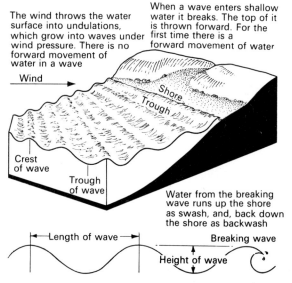

The wind throws the water surface into undulations, which grow into waves under wind pressure. There is no forward movement of water in a wave

When a wave enters shallow water it breaks. The top of it is thrown forward. For the first time there is a forward movement of water

Wind
Shore
Trough
Crest of wave
Trough of wave

Water from the breaking wave runs up the shore as swash, and, back down the shore as backwash

Length of wave
Breaking wave
Height of wave

Definition of Terms

Find out the meanings, of *coast*, *shore* and *beach* by studying the diagrams on the left. The highest water level refers to the level reached by the most powerful storm waves. The *coastline* is the margin of the land. This is also the *cliff line* on rocky coasts. The height and power of a wave depend upon the strength of the wind and the *fetch* (distance of open water over which it blows). The stronger the wind and the greater the fetch, the more powerful the wave. Storm waves are particularly powerful.

How waves are caused

Waves are caused by winds and the following diagrams show how this takes place, how a wave grows and the movement inside a wave.

1 In this diagram the wind, which is shown as four layers, blows over a sea surface. The surface exerts a frictional drag on the bottom layer, and this layer exerts a frictional drag on the layer above it, and so on. The top layer has the least drag exerted on it which means that the layers of air move forward at different speeds. The air tumbles forward and finally develops a circular motion. This motion exerts downward pressure (D.P.) on the surface at its front and an upward pressure (U.P.) at its rear. The surface begins to take on the form of a wave.

1
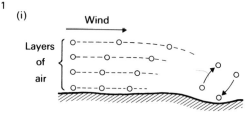
(i)
Wind
Layers of air
Surface of sea

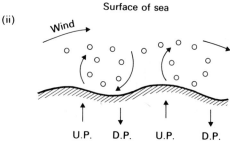
(ii)
Wind
U.P.   D.P.   U.P.   D.P.

2 This diagram shows the wind pressing on the back of a developing wave, thus causing it to steepen. The back of the wave tumbles forward but it moves back later and slows the forward movement of the front of the wave. This causes the wave to grow bigger.

2

(i)

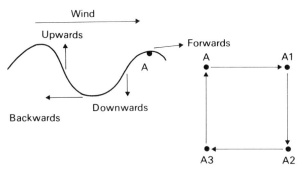

Back   Back

(ii)

(iii) *fully-formed wave*

Wind   Front   Back

3 This diagram shows the four component movements in a wave. Any particle of water at A moves to A1, to A2, to A3 and then back to A. It is therefore the wave form which moves and not the water.

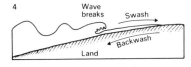

Wind

Upwards

Forwards
A   A1

A

Backwards   Downwards

A3   A2

4 When a wave enters shallow water its top falls forward and its water is thrown forward. The water thrown up the beach by breaking waves is called the *swash*. When the swash drains back down the beach it is called the *backwash*.

Wave breaks   Swash

Backwash

Land

4

Types of waves

The swash of a wave is always more powerful than the backwash because it has the force of the breaking wave behind it. When waves break at a rate of ten or less a minute, each breaking wave is able to run its course without interference from the wave behind it. These waves are called *constructive waves*. When waves break more frequently, especially over fifteen a minute, then the backwash of a wave runs into the swash of the wave behind. The force of the swash is therefore reduced in comparison with the force of the backwash. These waves remove pebbles and sand from a coast. They are *destructive waves*.

Waves as Agents of Erosion

Wave erosion consists of three parts:

1 *Corrasive action*: boulders, pebbles and sand are hurled against the base of a cliff by breaking waves and this causes undercutting and rock break-up.

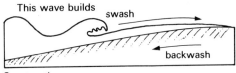

This wave builds   swash

backwash

Constructive wave
The swash is more powerful than the backwash.

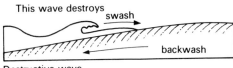

This wave destroys   swash

backwash

Destructive wave
The backwash is more powerful than the swash.

2 *Hydraulic action*: Water thrown against a cliff face by breaking waves causes air in cracks and crevices to become suddenly compressed. When the waves retreat the air expands; often explosively. This action causes the rocks to shatter as the cracks become enlarged and extended.

3 *Attrition*: Boulders and pebbles dashed against the shore are themselves broken into finer and finer particles.

Features Produced by Wave Erosion

Cliffs and Wave-cut Platforms

The Strandflat off the west coast of Norway is a good example of a wave-cut platform. This platform is over 50 kilometres (30 miles) wide.

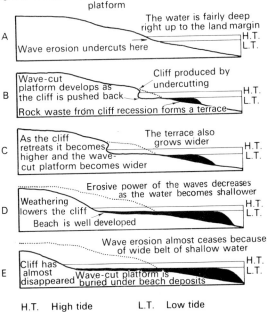

Stages in the development of a cliff and wave-cut platform

A   The water is fairly deep right up to the land margin
H.T.
L.T.
Wave erosion undercuts here

B   Wave-cut platform develops as the cliff is pushed back   Cliff produced by undercutting
H.T.
L.T.
Rock waste from cliff recession forms a terrace

C   As the cliff retreats it becomes higher and the wave-cut platform becomes wider   The terrace also grows wider
H.T.
L.T.

D   Erosive power of the waves decreases as the water becomes shallower
Weathering lowers the cliff
H.T.
L.T.
Beach is well developed

E   Wave erosion almost ceases because of wide belt of shallow water
Cliff has almost disappeared   Wave-cut platform is buried under beach deposits
H.T.
L.T.

H.T.   High tide   L.T.   Low tide

Types of Cliffs

The rocks of some cliffs are in layers which slope landwards (fig. A). In other cliffs the rock layers slope seawards and blocks of rock loosened by erosion easily fall into the sea. These cliffs are often very steep and overhanging (fig. B). Landslides are quite common on cliffs more especially on those formed of alternate layers of pervious and impervious rocks.

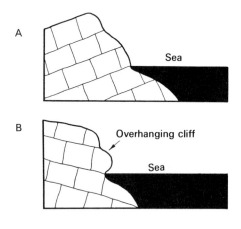

Part of the Devon Coast of S.W. England

THE ROCKS DIP STEEPLY LANDWARDS

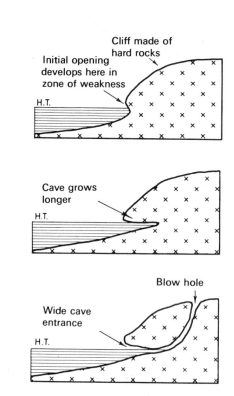

Cliff made of hard rocks

Initial opening develops here in zone of weakness

H.T.

Cave grows longer

H.T.

Blow hole

Wide cave entrance

H.T.

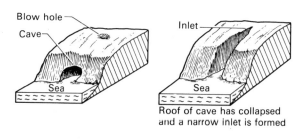

Blow hole

Cave

Sea

Inlet

Sea

Roof of cave has collapsed and a narrow inlet is formed

A Blow Hole in Scotland

Caves, Arches and Stacks

These are minor features produced by wave action during the process of cliff formation. A cave develops along a line of weakness at the base of a cliff which has been subjected to prolonged wave action. It is a cylindrical tunnel which extends into the cliff, following the line of weakness, and whose diameter decreases from the entrance. If a joint extends from the end of the tunnel to the top of the cliff, this becomes enlarged in time and finally opens out on the cliff top to form a *blow hole*. The roof of the cave ultimately collapses and a long narrow sea inlet forms.

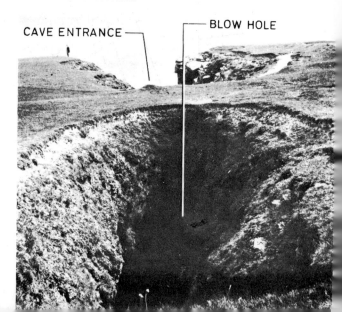

CAVE ENTRANCE

BLOW HOLE

Caves which develop on either side of a headland such that they ultimately join together, give rise to a natural *arch*. When the arch collapses, the end of the headland stands up as a *stack*. In time this is completely removed by wave erosion.

The diagrams below show the stages in the development of an arch and a stack.

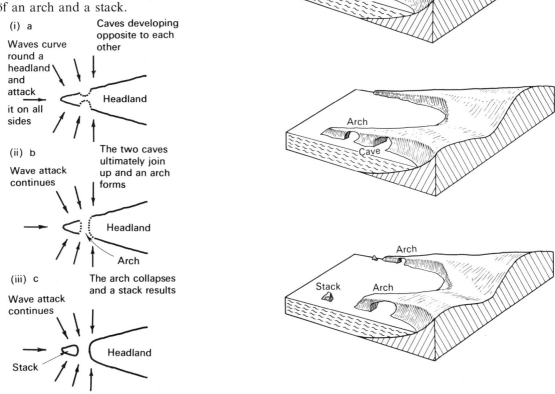

(i) a

Waves curve round a headland and attack it on all sides

Caves developing opposite to each other

Headland

(ii) b

Wave attack continues

The two caves ultimately join up and an arch forms

Headland

Arch

(iii) c

Wave attack continues

The arch collapses and a stack results

Stack

Headland

Natural stack and bay on the Gaspé Coast of Quebec (Canada)

Wave Transport

All the material carried by breaking waves is called the *load*. Some of this comes from rivers entering the sea, some from landslides on cliffs, and the rest comes from wave erosion. The load consists of mud, sand and shingle.

The swash and backwash push and drag material up and down the beach. When waves break obliquely to the shore the swash moves obliquely up the beach but the backwash runs back at right angles to the shore as shown in *fig. a*. You will see from this diagram that material is gradually carried along the beach by these two actions which together constitute *longshore drift*. The removal of material by longshore drift can be stopped by building groynes or walls out to sea (*fig. b*).

Material is also carried off-shore into deeper water by the *undertow*. This is an under current which balances the piling up of water along the coast by breaking waves and high tides (*fig. c*).

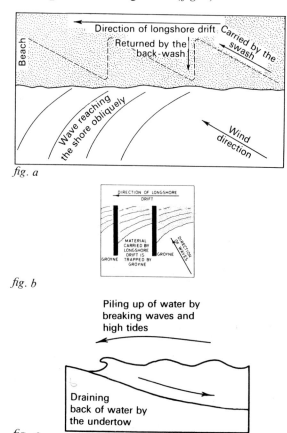

fig. a

fig. b

fig. c

Erosion is Dominant on a Highland Coast. Examine *fig. d*. The only depositional feature is a narrow beach.

Deposition is Dominant on a Lowland Coast. Examine *fig. e* and compare it with *fig. d*. Along the highland coast there is deep water close to the land.

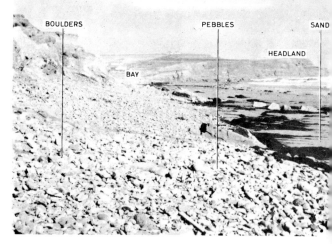

BOULDERS   PEBBLES   SAND   HEADLAND   BAY

Wave-deposited material on the Devon Coast, England

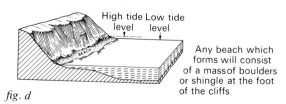

High tide level   Low tide level

Any beach which forms will consist of a mass of boulders or shingle at the foot of the cliffs

fig. d

The photograph above shows that the sea, like the wind and rivers, sorts its load on deposition. Moving down the beach the sequence of deposits is boulders, pebbles, sand and mud. The coast is gently sloping. This is shown by the wide expanse of beach.

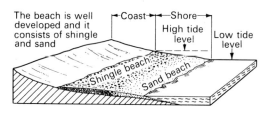

The beach is well developed and it consists of shingle and sand   Coast   Shore   High tide level   Low tide level   Shingle beach   Sand beach

fig. e

Wave Depositional Features

The chief of these are:
  (i) Beaches
 (ii) Spits and bars
(iii) Mud flats.

Beach: The main action of constructive waves is to deposit pebbles, sand and mud, which, when deposited along a coast, form a gently sloping platform, called a beach. The material of which a beach is composed is transported along a coast by longshore drift. Beaches usually lie between high and low water levels but storm waves along some coasts throw pebbles and stones well beyond the normal level reached by waves at high tide. The material deposited in this way produces a ridge called a *storm beach*.

Wave action in bays is usually not strong and deposition is the dominant action. Beaches called *bay-head beaches* develop at the heads of bays (*fig. f*). These beaches do not extend to the headlands where wave erosion is dominant. Good examples of beaches occur along the east coast and parts of the west coast of Peninsular Malaysia. A particularly fine example of bay-head beaches occurs between Port Dickson and Cape Rachado on the west coast of Peninsular Malaysia.

Bays and coves between
headlands develop bay beaches

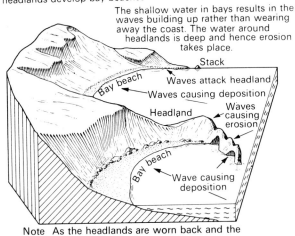

The shallow water in bays results in the waves building up rather than wearing away the coast. The water around headlands is deep and hence erosion takes place.

Stack
Waves attack headland
Waves causing deposition
Headland
Waves causing erosion
Bay beach
Bay beach
Wave causing deposition

Note   As the headlands are worn back and the bays are filled in the coast becomes straightened. This tends to smoothen out the coast rather than to indent it.

fig. f

Spit: Material which is eroded from a coast may be carried along the coast by longshore drift and deposited further along the coast as a spit. This is likely to happen along indented coasts and coasts broken by river mouths.

A spit is a low, narrow ridge of pebbles or sand joined to the land at one end with the other end terminating in the sea. A spit sometimes develops at a headland and projects across a bay. As waves swing into the bay obliquely, the end of the spit becomes curved or hooked (*fig. g*).

A spit across the entrance of a bay

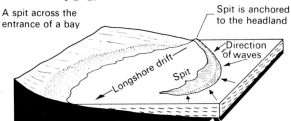

Spit is anchored to the headland
Direction of waves
Longshore drift
Spit

fig. g

When longshore drift operates across a river's mouth a zone of slack water develops between longshore drift and the river and any material carried by longshore drift is deposited. The deposited material forms a spit which may, in time, extend across the mouth of the river. When this happens the river's outlet may be diverted (*fig. h*) or the river's mouth may be converted into a *lagoon* (*fig. i*). A good example of a spit across a river's mouth occurs at the mouth of Sungai Kelantan (Peninsular Malaysia). A spit may also develop across a bay.

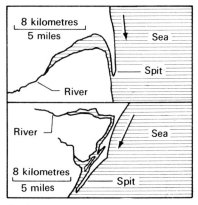

8 kilometres
5 miles
Sea
Spit
River

River
Sea
8 kilometres
5 miles
Spit

fig. h

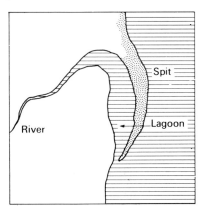

Spit
Lagoon
River

fig. i

Bar: This is very similar to a spit. A common type of bar is that which extends right across a bay. This starts as a spit growing out from a headland but ultimately it stretches across the bay to the next headland. This type of bar is called a *bay-bar*. Many bay-bars do have breaks in them where tidal

Chesil Beach in S.W. England

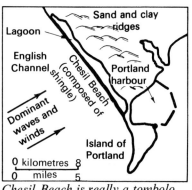

Chesil Beach is really a tombolo because it joins an island to the mainland.

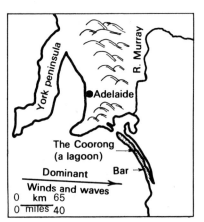

Coast of South Australia

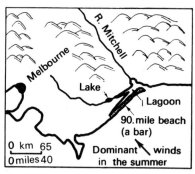

Coast of Victoria (Australia)

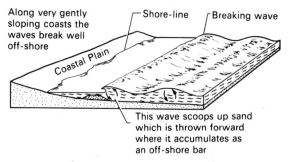

action prevents the bar from being continuous. Along the coast of Poland, bay-bars are called *nehrungs*. When a bar links an island to the mainland it is called a *tombolo*. Chesil Beach, in southern England, is a very good example.

Off-shore Bar: This develops only along very gently sloping coasts such as the southern part of the Atlantic Coast of North America. The two diagrams at the bottom of the page will help you to see how an off-shore bar can be built by wave action.

Mud flats: Tides tend to deposit fine silts along gently shelving coasts, especially in bays and estuaries. The deposition of these silts together, perhaps, with river alluvium, results in the building up of a platform of muds called a mud flat. Salt-tolerant plants soon begin to colonise the mud flat which in time becomes a swamp or marshland. In tropical regions, mud flats often become mangrove swamps. Mud flats are usually crossed by winding channels kept clear of vegetation by tidal action. At low tide these channels often contain little, if any, water.

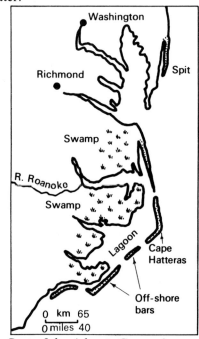

Part of the Atlantic Coast of North America

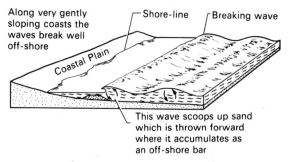

Along very gently sloping coasts the waves break well off-shore

This wave scoops up sand which is thrown forward where it accumulates as an off-shore bar

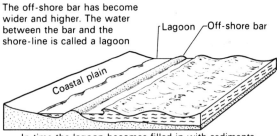

The off-shore bar has become wider and higher. The water between the bar and the shore-line is called a lagoon

In time the lagoon becomes filled in with sediments forming swamp or marsh and finally dry land

TYPES OF COASTS

Neither the level of the land nor the level of the sea remains unchanged for long periods of time. During the last Ice Age the level of the sea was lower than it is today because large quantities of water were locked up in the ice masses which covered extensive parts of Europe and North America. The Ice Age was followed by a gradual return to warmer conditions and the ice sheets melted and their waters returned to the sea. The level of the sea rose and some coastal regions were submerged. Land masses too change their level. Sometimes coastal regions were depressed; sometimes they were uplifted. After the disappearance of the ice sheets parts of the land masses were slowly uplifted through the removal of the great weight of ice.

It will be seen that coastal regions may be either submerged or uplifted by changes in land or sea levels. There are therefore two basic types of coast: *submerged* and *emerged*. Each can be sub-divided into highland and lowland types to give:

Submerged Coasts
1 Highland type
2 Lowland type

Emerged Coasts
1 Highland type
2 Lowland type

Submerged Highland Coasts
There are three main types:
1 Ria Coast
2 Longitudinal Coast
3 Fiord Coast

Ria Coast: When a highland coast is submerged the lower parts of its river valleys become flooded. These submerged parts of the valley are called rias. Such rias are common in S.W. Ireland, S.W. England, N.W. Spain, and Brittany.

Due to submergence the coast becomes indented and the tips of headlands may be turned into islands.

Longitudinal Coast: When a highland coast whose valleys are parallel to the coast is submerged, some of the valleys are flooded and the separating mountain ranges become chains of islands. These valleys are sometimes called *sounds*, e.g. Puget Sound in Washington (U.S.A.). This type of coast occurs in Yugoslavia and along parts of the Pacific coasts of North and South America.

Fiord Coast: When glaciated highland coasts become submerged the flooded lower parts of the valleys are called *fiords*. The three diagrams on page 82 show how the fiords have developed. During

Ria Coast

Before submergence

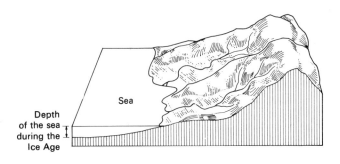

After submergence

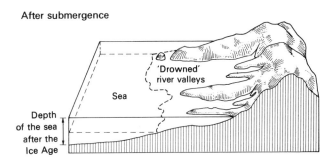

Longitudinal Coast

Before submergence

Mountain ranges and valleys are parallel to the coast

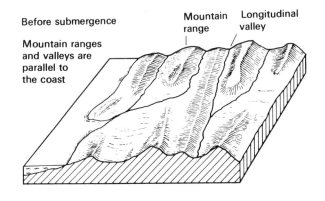

After submergence

Coastal mountain range has been turned into a chain of islands

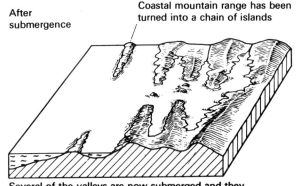

Several of the valleys are now submerged and they form long narrow inlets parallel to the coast

glaciation the river valleys become widened and deepened. After the glaciers have disappeared and the sea has risen the steep-sided valleys are 'drowned'. Notice that the water inside the fiord is much deeper than it is at the entrance of the fiord. Fiords have steeper sides and deeper water than rias. All the fiord coasts lie in the belt of prevailing westerly winds and are on the western sides of land masses. It was in these regions that vast amounts of snow and ice accumulated in the Ice Age. Some of the best examples of fiord coasts occur in Chile, South Island of New Zealand, Greenland, Norway and British Columbia.

Ria of the River Yealm on the South Devon Coast

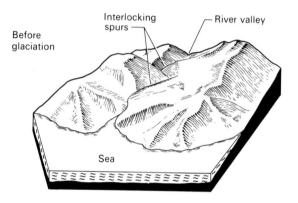

Before glaciation

Interlocking spurs — River valley

Sea

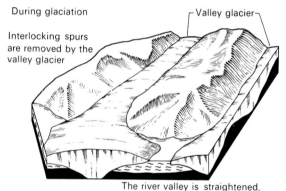

During glaciation

Valley glacier

Interlocking spurs are removed by the valley glacier

The river valley is straightened, widened and deepened (below sea level) by the glacier

Trollfiord in Northern Norway. Compare and contrast this with the photograph above

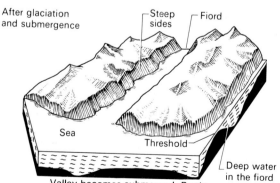

After glaciation and submergence

Steep sides — Fiord

Sea

Threshold

Deep water in the fiord

Valley becomes submerged. Partly caused by over deepening, and partly by a rise in sea level

The Value of Rias and Fiords to Man
1 Both rias and fiords often provide natural 'harbours'.
2 It is often extremely difficult to get inland from the head of a fiord because of the mountainous country. A fiord, therefore, is not very useful as a site for a port.
3 It is usually easy to get inland from the head of a ria and because of this a ria is sometimes the site of a port.
4 Settlement is difficult along the sides of a fiord because there is little or no level land. Fiord settlements occur at the head of a fiord where there is level land.

Wave Action Alters a Submerged Highland Coast

I. In the beginning

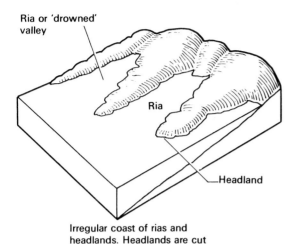

Ria or 'drowned' valley

Ria

Headland

Irregular coast of rias and headlands. Headlands are cut back by wave erosion, and cliffs, caves and stacks form

II. Stage of youth

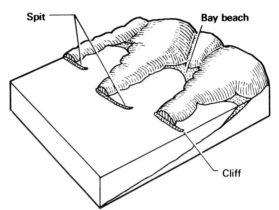

Spit

Bay beach

Cliff

Wave deposition is more important than erosion. Spits and bay beaches are formed. The coast is becoming straighter for erosion of the headlands still goes on

III. Stage of early maturity

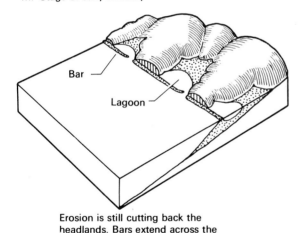

Bar

Lagoon

Erosion is still cutting back the headlands. Bars extend across the bays which are now turned into lagoons. These are being filled in with sediments and marshes form

IV. Stage of late maturity

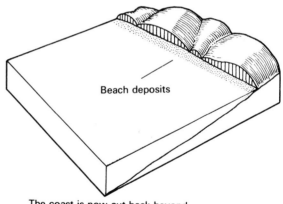

Beach deposits

The coast is now cut back beyond the heads of the bays, and it is now almost straight

Submerged Lowland Coast

A rise in sea level along a lowland coast causes the sea to penetrate inland along the river valleys, often to considerable distances. The flooded parts of the valleys are called *estuaries*. Marshes, swamps and mud flats can often be seen in estuaries at low tide. The Baltic coasts of Poland and Germany and the Dutch coast are good examples of estuarine coasts.

Emerged Highland Coast

An old sea beach backed by a sea cliff lying from 7·5 metres to 30 metres (25 to 100 feet) above sea level often characterises this type of coast. These two features could only have been produced by sea action, but since the sea no longer reaches them, it is evident that there has been a change in either sea level or the level of the land. Raised beaches are common in western Scotland.

OLD SEA CLIFFS

OLD SEA BEACH

PRESENT DAY SEA CLIFFS

PRESENT DAY BEACH

Raised Beach in Scotland

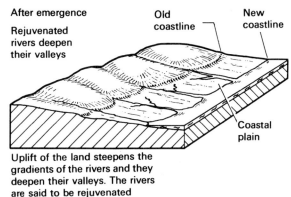

After emergence

Rejuvenated rivers deepen their valleys

Old coastline

New coastline

Coastal plain

Uplift of the land steepens the gradients of the rivers and they deepen their valleys. The rivers are said to be rejuvenated (made young again)

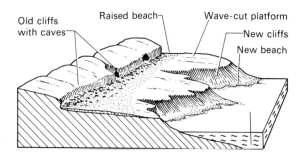

Old cliffs with caves

Raised beach

Wave-cut platform

New cliffs

New beach

Emerged Lowland Coast

This forms when a part of the continental shelf emerges from the sea and forms a coastal plain. The coast has no bays or headlands and deposition takes place in the shallow water off-shore, producing off-shore bars, lagoons, spits and beaches. Examples of this type occur along the south-east coast of the U.S.A. and the north coast of the Gulf of Mexico. The development of ports is difficult.

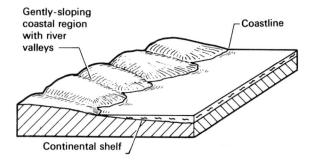

Before emergence

Gently-sloping coastal region with river valleys

Coastline

Continental shelf

EXERCISES

1 Briefly explain the meanings of the following terms:
  (a) swash
  (b) backwash
  (c) longshore drift
  (d) constructive wave
  (e) beach
  Choose any *three* features, and for *each*:
  (i) draw a well-labelled diagram to illustrate its main characteristics
  (ii) say how it develops
  (iii) state the type of coast where it occurs.

2 Choose *three* of the following physical features:
  (a) cliff
  (b) headland
  (c) stack
  (d) cave
  (e) blow hole
  For *each* feature
  (i) describe its appearance and mode of formation
  (ii) name and locate an area where an example may be seen.
  Illustrate your answer with relevant diagrams.

3 Describe the main destructive and constructive processes at work along coasts and make a list of some of the features which these processes produce. You should give specific examples where possible and your answer should be illustrated with well-labelled diagrams.

4 (a) By using well-annotated diagrams, state the main differences between the following coastlines:
    (i) ria coast
    (ii) fiord coast
    (iii) longitudinal coast
    (iv) estuarine coast
  (b) For *each* type of coastline, name *one* region where an example of it may be seen.

Objective Exercises

1 Which one of the following features is **not** produced by wave erosion?
   A headland
   B stack
   C beach
   D cliff
   E blow hole

   A B C D E
   □ □ □ □ □

2 The base of a cliff is undercut by rocks and sand being hurled at it by breaking waves. This process is called
   A attrition
   B corrasion
   C solution
   D hydraulic action
   E abrasion

   A B C D E
   □ □ □ □ □

3 Various coastal features are formed before a stack is finally produced. In what order do the features produced take place?
   A cave, headland, arch, stack
   B headland, cave, arch, stack
   C arch, cave, headland, stack
   D headland, arch, cave, stack
   E cave, arch, headland, stack

   A B C D E
   □ □ □ □ □

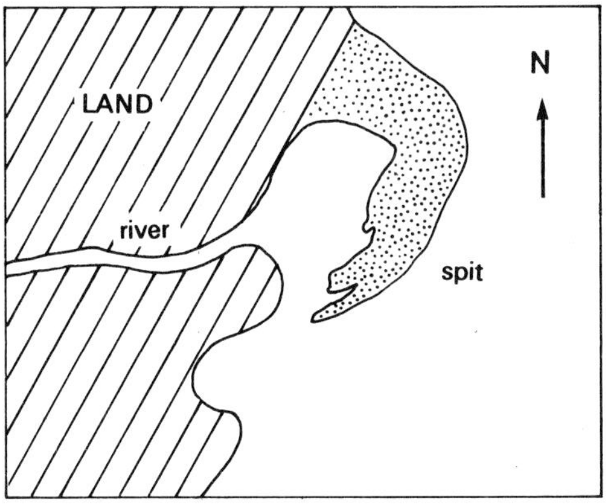

This is a diagram of a coastal area.

4 This direction of longshore drift as suggested by the spit is mainly
   A north to south
   B south to north
   C west to east
   D east to west
   E south-west to north-east

   A B C D E
   □ □ □ □ □

5 All of the following features are produced by wave deposition **except**
   A sand bar
   B beach
   C stack
   D spit
   E mud flat

   A B C D E
   □ □ □ □ □

6 Which one of the following features is **not** predominantly due to wave deposition?
   A off-shore bar
   B beach
   C mud flat
   D spit

   A B C D
   □ □ □ □

7 Which of the coasts listed below is predominantly a fiord coast
   A coast of Norway
   B western coast of South America
   C western coast of England
   D western coast of Australia

   A B C D
   □ □ □ □

8 A coastline which exhibits drowned river valleys is called a
   A fiord coast
   B submerged coast
   C ria coast
   D lowland coast

   A B C D
   □ □ □ □

9 Raised beaches usually occur along
   A emerged highland coasts
   B emerged lowland coasts
   C submerged highland coasts
   D submerged lowland coasts

   A B C D
   □ □ □ □

10 The two features marked '1' and '2' in the diagram below represent respectively
   A stack and cliff
   B headland and beach
   C arch and stack
   D headland and stack

   A B C D
   □ □ □ □

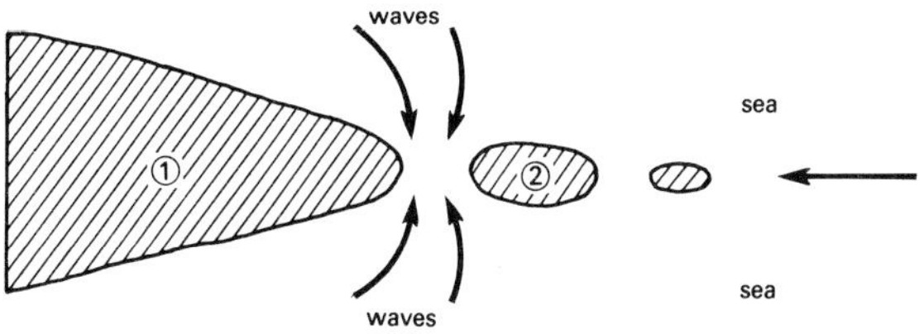

8   Coral Reefs and Islands

CORAL COASTS

Nature of coral

It is a limestone rock made up of the skeletons of tiny marine organisms called *coral polyps*. The tube-like skeletons in which the organisms live extend upwards and outwards as the old polyps die and new ones are born. Coral polyps cannot grow out of water and they have therefore formed below the level of low tide. Coral polyps thrive under these conditions:
1 sea temperature of about 21°C (70°F).
2 sunlit, clear salt water down to a depth of about 55 metres (180 feet).

Extensive coral formations develop between 30°N. and 30°S. especially on the eastern sides of land masses where warm currents flow near to the coasts. They do not develop on the western coasts in these latitudes because of the cool currents which flow along these coasts.

Types of Coral Formation

Coral masses are called *reefs* and there are three types:
1 *Fringing Reef*: a narrow coral platform separated from the coast by a lagoon which may disappear at low water.
2 *Barrier Reef*: a wide coral platform separated from the coast by a wide, deep lagoon.
3 *Atoll*: a circular coral reef which encloses a lagoon.

Structure of a Coral Reef

Most reefs are fairly narrow and the coral platform lies near to low water level. The seaward edge is steep and pieces of coral broken off by wave action are thrown up onto the platform where they form a low mound. On the landward side of this the breaking waves deposit sand in which the seeds of plants, such as coconut, readily germinate. Coral atolls in the Pacific are of this type.

Section across a Coral Reef

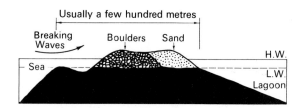

Fringing Reef

This type of reef consists of a platform of coral which is connected to, and which is built out from, a coast. The surface of the platform is usually flat or slightly concave and its outer edge drops away steeply to the surrounding sea floor. A shallow lagoon usually occurs between the coast and the outer edge of the reef.

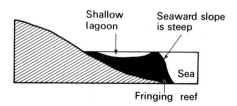

Barrier Reef

This reef is similar to a fringing reef except that it is situated several miles off the coast and is separated from it by a deep water lagoon. The coral of a barrier reef is often joined to the coast although the lagoon may be too deep for coral to grow on its bed. The origin of reefs is discussed later and this offers a possible explanation to this phenomenon. Some barrier reefs lie off the coasts of continents, e.g. Great Barrier Reef along the east coast of Australia. Others occur around islands forming a continuous reef except for openings on the leeward side. In some cases, fringing reefs develop on the inner side of lagoons which lie between a barrier reef and the coast of the island that it encircles.

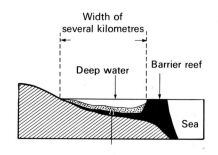

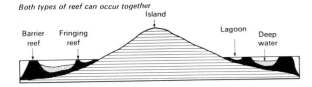

Both types of reef can occur together

Atoll

The diagrams below and on the right show drawings of a typical atoll. Atolls are particularly common in the Pacific and Indian Oceans. Some atolls are very large, e.g. that of Suvadiva in the Maldives. The lagoon of this atoll is 64 kilometres (40 miles) across and its reef extends for 190 kilometres (120 miles). The Gilbert and Ellice Islands of the Pacific are all atolls.

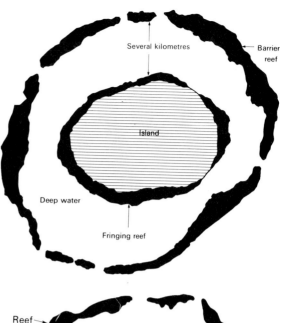

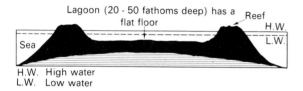

An Atoll in the Pacific Ocean

The Origin of Coral Reefs

Many theories have been put forward to explain the origin of reefs, and in common with so many theories explaining the possible origin of other landforms, no one can say which is the true one.

All that we can do is to examine these theories and accept that which sounds the most reasonable.

Theory I (Daly's Theory). This is based on the changing level of the sea during and after the last Ice Age.

Fringing Reef   *Barrier Reef*   *Atoll*

Before the Ice Age

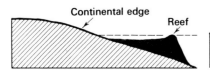

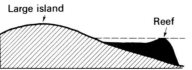

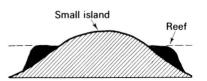

During the Ice Age

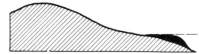

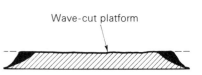

As sea level fell wave erosion removed most of the reef.

As sea level fell wave erosion cut a terrace into the reef and the edge of the island.

As sea level fell wave erosion removed the top of the island and turned it into a wave-cut platform.

After the Ice Age

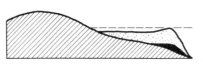

New reef

New reef

A rising sea level plus a return to warmer conditions caused the reef to grow up and form a barrier reef.

A rising sea level plus a return to warmer conditions caused the reef to grow up again.

A rising sea level plus a return to warmer conditions caused the reef to grow up again and form an atoll.

Theory II (Darwin's Theory). This depends on the subsidence of land masses.

As an island subsides the coral reef grows upwards and outwards keeping pace with the subsidence.

According to this theory fringing reefs pass into barrier reefs which in turn pass into atolls. This theory was first suggested by Charles Darwin. The two theories mentioned above can only be used with reference to coral island reefs.

Fringing Reef

Barrier Reef

Atoll

Coral broken from the reef by the waves is deposited inside the reef where it forms the floor of the lagoon

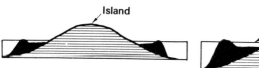

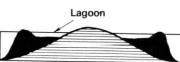

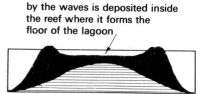

EXERCISES

1 Select *two* of the following coral features: atoll, fringing reef and barrier reef, and for *each* feature selected:
   (i) briefly explain how it may have originated and describe its characteristics
   (ii) name *one* example, or *one* region where an example may be seen
   (iii) draw a well-labelled diagram to illustrate its appearance.
2 Choose *two* of the following physical features: a volcano, a barrier reef, a delta and a raised beach, and for *each* feature chosen:
   (i) describe its appearance
   (ii) draw an annotated diagram to show its main characteristics
   (iii) name *one* example, or *one* region where an example may be seen
   (iv) suggest how it was formed.

Objective Exercises

1 What is the name given to an almost circular coral reef inside which is a lagoon?
   A fringing reef
   B barrier reef
   C atoll
   D coral island

   A B C D
   □ □ □ □

2 Which one of the following conditions is **not** important for the growth of coral?
   A wave-free salt water
   B clean salt water
   C warm seas (about 21°C)
   D plenty of sunlight

   A B C D
   □ □ □ □

9 Features Produced by Glaciers

ICE ACTION AND THE FEATURES IT PRODUCES

When the temperature of the air falls below 0°C (32°F) some of the water vapour condenses and freezes into ice crystals which fall to earth as *snow*. Many regions in the high latitudes receive snow in the winter season but in most of them the snow melts in the following summer. If some of it fails to melt then a perpetual cover of snow results. This happens in Greenland, Antarctica and on the tops of some high mountains. The level above which there is a perpetual snow cover is called the *snowline*. The height of this ranges from sea level around the Poles to 4800 metres (15 700 feet) in the mountains of East Africa which are on the equator. When the accumulation of snow in a region increases from year to year it gradually turns into ice by its own weight. About 1 000 000 years ago the climates of regions in the high latitudes began to get colder and colder and not all of the winter snowfall melted in the following summers. The accumulations of snow increased in area and in depth in the polar regions, in the northern part of North America and the north-western part of Europe. The snow of these vast *snow fields* gradually turned into ice which extended over most of the lowlands and some of the mountains. Masses of ice which cover large areas of a continent are called *ice sheets*, and those which occupy mountain valleys are called *valley glaciers*. Today ice sheets occur in Antarctica and Greenland, and valley glaciers in the Himalayas, Andes, Alps and Rockies. The period when the high latitudes were buried beneath ice sheets is known as the *Ice Age*. With the return to warmer conditions most of the ice melted. However, there are still extensive regions around the Poles and smaller areas in the mountain systems named above which still have glaciers, and these regions are therefore still in the Ice Age.

Ice action greatly changes the appearance of a region. Highlands are subjected to erosion and lowlands to deposition. In many parts of the northern continents which are now free from ice, striking features of both glacial erosion and deposition can be clearly seen. With the melting of the ice at the end of the Ice Age enormous quantities of water were set free. Some of this collected in hollows or was held back by glacial deposits and formed lakes. The Great Lakes of North America and the lakes of Finland were formed in this way. Most of the melt waters, however, flowed as rivers into the sea.

These rivers carried large quantities of morainic deposits which were later spread over the land outside the regions which lay under the ice. Here the deposits formed extensive plains called *outwash plains*. These are usually very sandy.

GLACIAL EROSION AND THE FEATURES IT PRODUCES

Glacial Erosion. Consists of two processes: (i) *plucking* (the tearing away of blocks of rock which have become frozen into the base and sides of a glacier), and (ii) *abrasion* (the wearing away of rocks beneath a glacier by the scouring action of the rocks embedded in the glacier).

Erosional Features. The most important of these are: U-shaped valley; hanging valley; cirque (corrie); arête; pyramidal peak. These features are chiefly produced by valley glaciers. Study diagram on page 91.

VALLEY GLACIERS

Snow falling in mountainous regions accumulates in depressions on mountain slopes which are not facing the sun. As the snow accumulation increases, a large part of it is turned to ice and there comes a time when there is more ice and snow than the depression can hold and some of it moves downslope to lower levels.

The movement of ice in the depression causes considerable erosion on the floor and on the sides of the depression. Eventually the depression is turned into a deep hollow. Erosion on the depression's floor is achieved by abrasion of the heavy accumulation of boulders embedded in the base of the ice mass. The floor becomes concave and the edge of the depression becomes ridge-like. This is called the *lip*. On the sides and back of the depression a different type of erosion takes place. The ice becomes frozen to the sides, especially to the back wall and when the ice moves forward it pulls pieces of rock out of the wall. This action is called plucking and it results in the back wall becoming very steep. When fully formed the hollow has the appearance of an arm chair and it is called a *corrie* or *cirque*.

Sometimes corries develop on adjacent mountain slopes and when fully formed only a knife-edge ridge, called an *arête* separates them. If corries develop on all sides of a mountain then the top of the mountain is reduced to a jagged peak, called

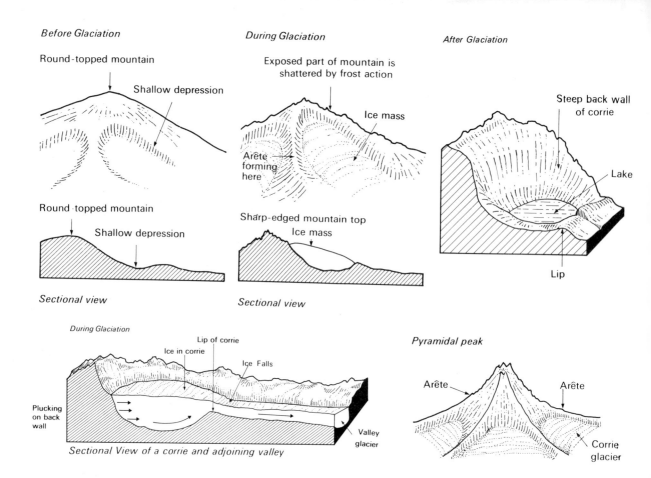

Before Glaciation

Round-topped mountain

Shallow depression

Round-topped mountain

Shallow depression

Sectional view

During Glaciation

Exposed part of mountain is shattered by frost action

Ice mass

Arête forming here

Sharp-edged mountain top

Ice mass

Sectional view

After Glaciation

Steep back wall of corrie

Lake

Lip

During Glaciation

Ice in corrie

Lip of corrie

Ice Falls

Plucking on back wall

Valley glacier

Sectional View of a corrie and adjoining valley

Pyramidal peak

Arête

Arête

Corrie glacier

a *pyramidal peak*, by the steepening of the back walls of the corries. The arêtes between the corries are pronounced and these, together with the pyramidal peak, are sharpened by frost action.

Ultimately, climatic changes may cause the ice to melt and disappear. When this happens a deep lake or *tarn* usually occupies a corrie. This overflows the lip, into the valley below, giving rise to a waterfall. A corrie glacier often extends down the valley as a valley glacier. As the glacier passes over the lip of the corrie, it becomes broken by vertical cracks. called *crevasses* which produce a series of ice falls on the surface. As it moves down the valley large quantities of boulders and rocks, produced by weathering, especially frost action, fall onto its surface and work their way between the valley sides and the glacier, and the valley bottom and the glacier. It is this material which gives a glacier its powers of erosion. Any material carried along by a glacier is called *moraine*.

How a glacier moves

If the accumulation of snow above the snowline increases from year to year, then a glacier will form. This will expand, and in time, extend below the snowline as a valley glacier in mountain regions. The glacier will continue to move down the valley as more ice accumulates above the snowline. Eventually a glacier will reach a point in the valley where the temperature is high enough to cause ice to melt and the glacier usually does not extend down-valley beyond this point. If the melting of the ice at the glacier *snout* (front) is balanced by the addition of new ice at the head of the glacier, then the glacier remains stationary. If ice is added faster than it melts, then the glacier will gradually extend farther down the valley. But when more ice melts than is added, the glacier will slowly retreat up the valley. These changes in the length of a glacier suggest a movement of a glacier in its valley. But the movement is not of the glacier as a whole. The pressure at the bottom, at the sides, and in the middle of a glacier is very great and this causes some of the ice to melt. The water so formed is turned back to ice by the intense cold very quickly but there is just sufficient time for the water to trickle a short distance. Throughout a glacier bits of ice are melting, trickling down-valley and then turning back to ice the whole time. This means that within the glacier there is a gradual down-valley movement.

The Rhône Glacier in Switzerland

The glacier is fed by the snow field which occupies the upper basin. Try to locate an ice fall, an arête, a cirque and a pyramidal peak.

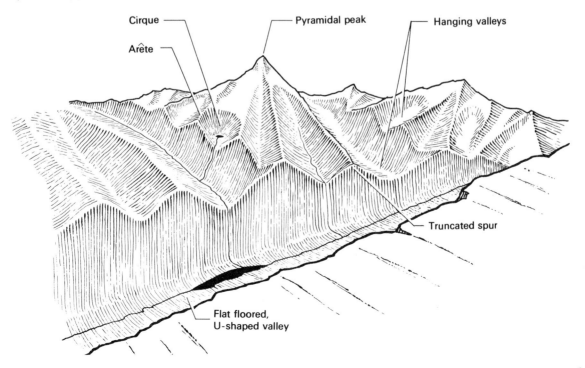

If a glacier extending down a valley enters a part of the valley which is wider than the rest, the ice of the glacier spreads out to fill the valley. This causes the upper layers of ice of the glacier to crack along lines parallel to the valley sides. These cracks are often very deep. Similar cracks develop in a glacier's surface when it passes over a part of the valley floor which is steeper than the rest. These cracks develop at right angles to the valley sides. All of these cracks are called *crevasses*.

How a glacier shapes a valley

As the amount of ice in a valley increases by addition of ice at the rear and by tributary glaciers entering the main valley, the power to erode by a valley glacier also increases. This results in a glacier deepening, straightening and widening a river valley. It cuts back the ends of spurs turning them into *truncated spurs*. One of the most noticeable effects of valley glacier erosion is the over-deepening of the valley which gives it a characteristic U shape.

U-shaped Valley

These diagrams show how a glaciated valley may have formed. Inter-valley divides are sharpened to give arêtes and pyramidal peaks. Examine the two contour maps and compare these with their respective block diagrams.

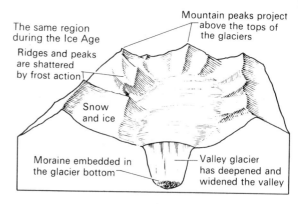

The same region during the Ice Age

Mountain peaks project above the tops of the glaciers

Ridges and peaks are shattered by frost action

Snow and ice

Moraine embedded in the glacier bottom

Valley glacier has deepened and widened the valley

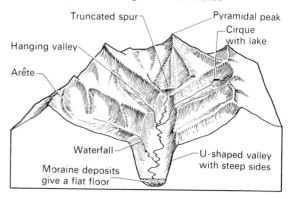

The same region after the glaciers have melted

Truncated spur

Pyramidal peak

Cirque with lake

Hanging valley

Arête

Waterfall

Moraine deposits give a flat floor

U-shaped valley with steep sides

U-shaped Valley

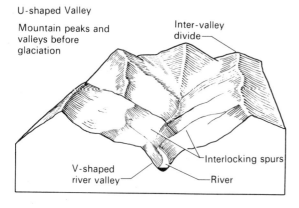

Mountain peaks and valleys before glaciation

Inter-valley divide

V-shaped river valley

Interlocking spurs

River

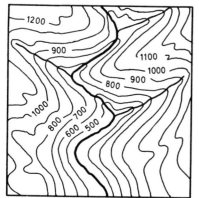

Hanging valleys

Vertical erosion of the main valley is much greater than that of the tributary valleys which contain many small glaciers or no glaciers at all. After the glaciers have retreated the floor of the main valley lies far below the floors of the tributary valleys which, as a result are called *hanging valleys*. Rivers and streams which usually take over the valleys after the glaciers have disappeared, often appear to be too small for the valleys. The streams of hanging valleys join the main river via waterfalls, which may be several hundred metres high. Waterfalls from hanging valleys sometimes build up alluvial fans of coarse materials.

The Lauterbrunnen Valley in Switzerland

Identify four important features in this valley which have been produced by glacial action.

Alluvial fan in a lake in Switzerland

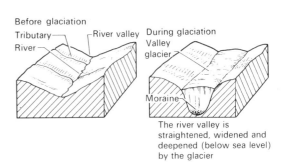

Before glaciation

Tributary — — River valley During glaciation

River Valley glacier

Moraine

The river valley is straightened, widened and deepened (below sea level) by the glacier

After glaciation

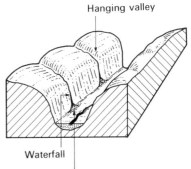

Hanging valley

Waterfall

Alluvial fan (see photograph)

Effects of Glaciers on Mountain Peaks and Ridges

We have seen that mountain peaks and ridges which project above the surface of glaciers and snow fields are shattered by frost action to give sharp-edged features. Sometimes, however, whole mountain regions are completely buried by glaciers. When this happens, there is virtually no frost action and the rugged peaks and jagged slopes are smoothed off as shown in the following diagrams. Compare these mountains with those of the diagram at the bottom left of page 91 which projected above the level of the glaciers and which were shattered by frost.

The Dent Blanche in Switzerland. Mt. Bernina and Mt. Roseg

This photograph shows a pyramidal peak, cirques and arêtes. The cirques contain snow fields. The glacier of the cirque in the foreground forms ice falls as it enters the valley below the cirque. Notice the sharpness of the arête on the right.

The Mourne Mountains in Ireland – compare with the photograph above

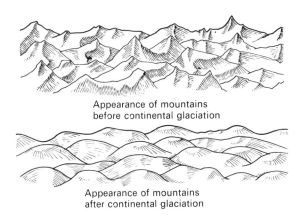

Appearance of mountains
before continental glaciation

Appearance of mountains
after continental glaciation

GLACIAL DEPOSITION AND THE FEATURES IT PRODUCES

A valley glacier carries large quantities of rock waste called *moraine*. Some of this is torn by the glacier from the bottom and sides of the valley and becomes embedded in the glacier. The rest falls onto the glacier surface from the mountain slopes where it has accumulated under frost action. The moraine which forms along the sides of a glacier is called *lateral moraine*; that along the front of a glacier is called *terminal moraine* and that at the bottom of a glacier is called *ground moraine*. When two glaciers join together, their inner lateral moraines coalesce (or join together) to give a *medial moraine*.

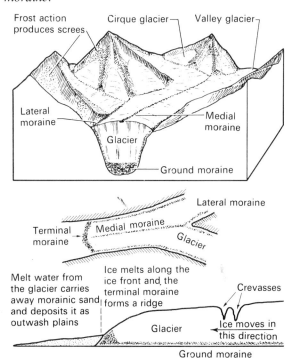

A terminal moraine only builds up when a glacier is stationary, that is, when the amount of ice being added at a glacier's snout is balanced by the amount

Nabesna Glacier in Alaska

of ice which is melting there. Some terminal moraines are very impressive. Material from a terminal moraine is always carried down-valley by the melt waters issuing from a glacier's snout, and is deposited as a layer called an *outwash plain*.

These types of moraine can often be clearly seen in glaciated valleys long after the glaciers have retreated.

Examine the photograph of the Nabesna Glacier in Alaska and see how many of these features you can recognise:
  (i) medial moraine
  (ii) truncated spur
  (iii) hanging valley
  (iv) cirque
  (v) pyramidal peak
  (vi) arête
  (vii) snow field
  (viii) lateral moraine.

ICE SHEETS

Ice sheets tend to move more slowly than valley glaciers but the erosion they cause can give rise to spectacular features depending on the structure and topography of the region.

Features produced by erosion

The ancient shields of Finland and northern Canada are two regions which were glaciated by ice sheets. In both regions hills were rounded and worn smooth and hollows were excavated in outcrops of soft rocks. These hollows are now the sites of lakes. Outcrops of harder rocks were smoothed on the side facing the ice to give a gentle slope, and plucked on the side facing away from the ice to give a steep

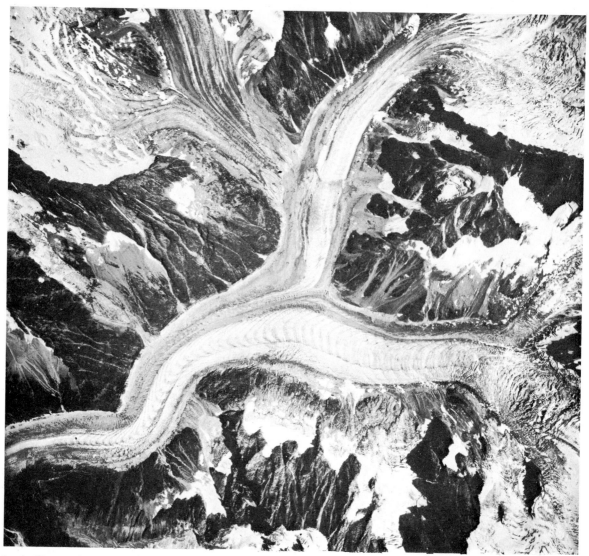

Mer de Glace in the French Alps

This glacier is 5½ kilometres (3½ miles) long and at its end the melt waters give rise to a river. How many glacier features can you recognise on this photograph?

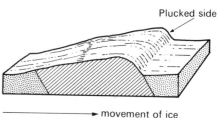

Plucked side

→ movement of ice

Roche moutonnée

After Glaciation

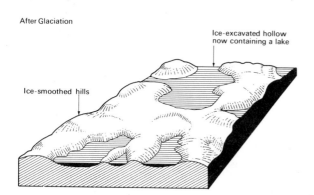

Ice-excavated hollow now containing a lake

Ice-smoothed hills

slope. Because of their characteristic shape they are called *roches moutonnees*. Outcrops of harder rocks, especially volcanic rocks were also shaped by ice erosion to give features similar to roches moutonnées, except that the sides facing the ice was steepened. The outcrop protected the softer rocks on the down-stream side from being eroded. These features, which are called *crag and tail*, have a gently sloping down-stream side.

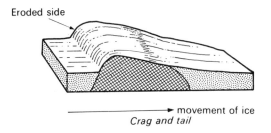

Eroded side

movement of ice
Crag and tail

Features produced by deposition

One of the most conspicuous features of lowlands which have been glaciated by ice sheets is the widespread morainic deposits. Thick deposits of clay containing angular rock particles, cover large parts of glaciated lowlands. Because of the numerous boulders in the clay, these deposits are called *boulder clay* deposits. The deposits are sometimes several hundred metres thick and their surface is marked by long rounded hills, called *drumlins*. Large blocks of rock, which are of a material quite different to that of the rocks of the region, often occur in regions which lay under ice sheets. These blocks are known as *erratics*, and they were obviously uprooted in one region and deposited in another.

This is what an ancient shield looks like after it has been glaciated. Rock hollows containing lakes, eskers and terminal moraines, often covered with pine trees, characterise the landscape.

The Lake Plateau of Finland

The edge of a boulder clay deposit is sometimes marked by a terminal moraine which will be well developed if the ice remained stationary for a long time. Glacial deposits quite different to boulder clay often occur on the outside of the terminal moraine. They are well-sorted, that is, coarse materials such as gravels occur immediately outside the moraine while finer deposits, such as sand, occur further away. Such deposits have been carried away from the ice sheets by melt waters emerging from the ice front. The deposits are water-sorted and build up outwash plains.

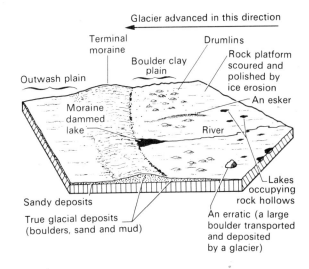

Rivers and streams occur inside most glaciers and these are heavily loaded with rock debris. As an ice front retreats the rivers build up ridge-like deposits called *eskers*. They develop on top of the boulder clay deposits.

Boulder clay plains, drumlins, terminal moraines, outwash plains and eskers rarely occur in such an ordered manner as has perhaps been suggested. In many regions ice sheets advanced, later retreated and then advanced again thus mixing up and sometimes obliterating the originally deposited materials. Also, since the final retreat of the ice, river erosion and weathering have modified and sometimes partially covered up or removed many of the glacial depositional features.

Note Valley glaciers also produce roche moutonnées, drumlins, eskers, etc.

The Value of Glaciated Regions to Man

I *Glacial Features of Value to Man*
  (i) Boulder clay plains are sometimes very fertile, e.g. East Anglia in Great Britain and parts of the Dairy Belt of North America.
  (ii) Old glacial lake beds are invariably fertile. Extensive areas of the Canadian Prairies producing vast amounts of wheat each year owe their prosperity to the rich alluviums which once collected on the floors of glacial lakes.
  (iii) Some glacial lakes, e.g. the Great Lakes of North America, are of real value as natural routeways.
  (iv) Waterfalls issuing from hanging valleys are sometimes suitable for the development of hydro-electric power. Both Norway and Switzerland develop large amounts of H.E.P. from such waterfalls.
  (v) Glaciated mountain regions attract tourists, especially during the winter season when heavy snowfalls make skiing and other sports possible.
  (vi) Some glacial lakes have cut deep overflow channels where they have drained away. Some of these channels today form excellent routeways across difficult country, e.g. the Hudson – Mohawk Gap which leads down to New York.
  (vii) Many glaciated valleys have benches or 'alps' high up on their sides. During the summer these 'alps' have good pasture, but during the winter they are covered with snow. Cattle are grazed on the alpine pastures during the summer and are brought down to the sheltered valley bottom pastures during the winter. This movement of animals is called *transhumance*, and it goes on in Switzerland, Norway and other mountainous countries.

II *Glacial Features of Little Value to Man*
Now let us look at the disadvantageous aspects of glaciation.
  (i) Boulder clay deposits in some regions, e.g. Central Ireland, have produced a marshy landscape which is of little or no value to agriculture.
  (ii) Many outwash plains contain infertile sands which give rise to extensive areas of waste land. It is true that this is sometimes of recreational value but from an agricultural standpoint such regions are negative.
  (iii) Extensive areas of land are sometimes turned into myriads of lakes by morainic deposits. Such lake landscapes offer little scope for development by Man.

EXERCISES

1 The following features often occur in glaciated regions: cirque (corrie), moraine, hanging valley, pyramidal peak, arête. Choose *three* of these features and for *each*,
  (i) briefly explain how it may have originated
  (ii) show its appearance by means of a well-labelled diagram
  (iii) name *one* region where an example may be found.

2 With the aid of diagrams, describe *three* of the following and explain how they may have been formed: esker, drumlin, terminal moraine, crag and tail, roche mountonnée.

3 (a) Outline the main differences between continental ice sheets and valley glaciers.
  (b) Briefly explain the main differences between ice action in mountain regions and ice action in lowland regions and name the characteristic physical features produced in each region.

4 By using well-labelled diagrams, explain the main differences between *three* of the following pairs of features:
  (i) truncated spur and interlocking spur
  (ii) crevasse and ice fall
  (iii) terminal moraine and boulder clay
  (iv) ice sheet and valley glacier
  (v) glaciated valley and river valley.

5 Briefly explain *three* of the following:
  (i) the sides of a valley glacier move more slowly than its middle
  (ii) boulder clay differs from an outwash plain in that its material is unsorted
  (iii) glaciated valleys are distinctly U-shaped
  (iv) the floor of a corrie is usually concave
  (v) tributary valleys usually join a glaciated valley high above the floor of the glaciated valley.

6 Choose *two* of the following landforms: a rift valley, an atoll, a glaciated valley, a fiord, and for *each* of the two chosen:
  (i) describe its characteristic features
  (ii) briefly explain how it has originated
  (iii) show the landform by means of an annotated diagram
  (iv) name *one* region where the landform may be seen.

Objective Exercises

1 The erosive action of a valley glacier depends most upon
  A the gradient and width of its valley
  B the width and thickness of the ice
  C the height of the glacier above sea level
  D the length of the glacier
  E the rate of movement of the glacier
    A B C D E
    □ □ □ □ □

2 Crevasses, which are cracks in the surfaces of glaciers, are produced by
  A the glacier moving over level land
  B the melting of the ice
  C differential movement in the ice
  D the thickness of the ice
  E rain action on the surface of the glacier
    A B C D E
    □ □ □ □ □

3 Which one of the following erosional features proves that ice can move uphill?
  A cirque
  B arête
  C truncated spur
  D alp
  E ice fall
    A B C D E
    □ □ □ □ □

4 A region which has not been glaciated may show
  A U-shaped valleys
  B corries in the mountains
  C waterfalls rushing down the steep sides of valleys
  D ox-bow lakes
  E hanging valleys
    A B C D E
    □ □ □ □ □

5 Which one of the following is most associated with ice action?
  A deltaic plain
  B loess deposits
  C outwash plains
  D tombolo
  E laterite plain
    A B C D E
    □ □ □ □ □

6 The first requirement for the formation of a glaciated U-shaped valley is
  A a glacier
  B heavy and continuous falls of snow
  C a mountainous terrain
  D a river valley
    A B C D
    □ □ □ □

7 All of the following features are produced by glacial deposition **except**
  A drumlin
  B moraine
  C esker
  D roche moutonnée
    A B C D
    □ □ □ □

8 Of the several features produced by the action of ice, some are the product of both erosion and deposition. Which of the following features is of this type?
  A arête
  B hanging valley
  C erratic
  D roche moutonnée
    A B C D
    □ □ □ □

9 This is a diagram of a crag and tail. In which direction did the ice predominantly move?
  A east to west
  B west to east
  C south to north
  D north to south
    A B C D
    □ □ □ □

10 The diagram below represents a glacier emerging from a valley near the foot of the mountains. The end of the glacier has melted without it retreating, and has deposited a feature at X which is called a
  A medial moraine
  B boulder clay
  C terminal moraine
  D drumlin
    A B C D
    □ □ □ □

10   The Oceans

The oceans and seas together cover about 70 per cent of the world's surface. There are five oceans and all are joined with one another. These are: Southern Ocean; Indian Ocean; Pacific Ocean; Arctic Ocean; Atlantic Ocean.

NATURE OF THE OCEAN FLOOR

Ocean floors, like continental surfaces, have relief features the chief of which are shown in the diagram. The edges of continents slope gently downwards under the surrounding oceanic waters. This part of the ocean floor is called the *continental shelf*. Along some coasts it is so narrow as to be almost absent. The best developed continental shelves are shown in the diagrams on page 101. Seawards of the shelf the ocean floor slopes more steeply, and this part of the floor forms the *continental slope*. The bed of the ocean sometimes rises up to give *ridges* some of which may appear above the surface of the ocean as *oceanic islands*. Below the ocean bed there are troughs and basins which are known as *deeps* or *trenches*. All the slopes shown in the diagram below are greatly exaggerated but the depths given are true.

Note Do not confuse *oceanic islands* with *continental islands*. The latter rise from the continental shelf.

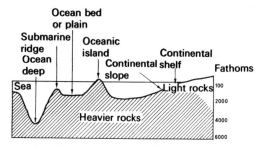

Generalised section across an ocean floor

Continental Shelves of the World
Northern Hemisphere
Eurasia
1 Along the coast of N.W. Europe
2 Along the coast of Siberia
3 The floor of the Yellow Sea
4 The floor of the Java Sea and the southern part of the S. China Sea

North America
1 Along the north-east coast of North America
2 The floor of Hudson Bay
3 Along the Gulf Coast of North America

Southern Hemisphere

Australasia
1 The floor of the Gulf of Carpentaria
2 The floor of the Australian Bight

South America
Along the coast of Patagonia

Africa
Very poorly developed

Value to Man
1 Sunlight easily penetrates the seas on continental shelves, and therefore there is an abundance of *plankton* or small green marine plants which usually results in an abundance of fish.
2 They increase the height of tides thus improving shipping facilities.

Ocean Deeps and Ridges (they all lie below 4000 metres)

Notice how the deeps flank the east coast of Asia, and compare the location of the deeps with the location of Young Fold Mountains (diagram, page 26) and of volcanoes (diagram top of page 37).

Deeps

Pacific Ocean
1 Mariana Trench (12 000 metres approx.)
          (39 240 feet approx.)
2 Philippine Trench (11 980 metres approx.)
          (39 175 feet approx.)
3 Tonga Trench (10 300 metres approx.)
          (33 680 feet approx.)
4 Japanese Trench (9300 metres approx.)
          (30 400 feet approx.)
5 Aleutian Trench (8380 metres approx.)
          (27 400 feet approx.)

Atlantic Ocean
1 Puerto Rico Deep (9625 metres approx.)
          (31 475 feet approx.)
2 Romanche Deep (8060 metres approx.)
          (26 355 feet approx.)
3 South Sandwich Trench (9090 metres approx.)
          (29 725 feet approx.)

Indian Ocean
Sunda Trench (8140 metres approx.)
          (26 620 feet approx.)

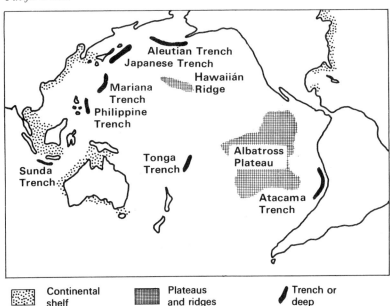

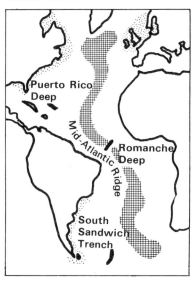

▨ Continental shelf	▨ Plateaus and ridges	◢ Trench or deep

Ridges
Pacific Ocean
1 Hawaiian Ridge
2 Albatross Plateau

Atlantic Ocean
Mid-Atlantic Ridge (arising from this are the oceanic islands of the Azores, Tristan da Cunha and Ascension).

Indian Ocean
A ridge extending from S. India to Antarctica.

THE NATURE OF SALT WATER
Salinity
Sea water contains mineral salts. Two important salts are sodium chloride (NaCl) and calcium bicarbonate $(CA(HCO_3)_2)$. The latter provides marine organisms with calcium carbonate $(CaCO_3)$ which is necessary to the formation of shells and bones. When water evaporates the salts are left behind. The saltiness of a sea depends upon the amount of evaporation taking place from its surface, and the amount of fresh water brought into it by rivers. Semi-enclosed seas do not mix freely with the oceans, so it is there that most variation is found. The Mediterranean and Red Seas are more salty than the oceans because they are located in regions of high temperatures and few rivers discharge into them. The Black and Baltic Seas are much fresher than the oceans. These seas are located in regions of low temperatures and fresh water is brought into them by rivers and melting ice. Inland seas like the Great Salt Lake (N. America) and the Dead Sea (Jordan) are very salty indeed.

Temperature
Water is heated by the sun's rays much more slowly than is land. It also loses heat to the air more slowly than does land. This causes the temperature of sea water to vary only slightly from season to season. The temperature of surface sea water ranges from about $-2°C$ (28°F) in polar regions to 26°C (about 79°F) in equatorial regions. The bottom water of the oceans is always cold, the temperature being about 1°C (about 34°F).

WATER MOVEMENTS IN THE OCEANS

There are two types of movement:

1 *Horizontal*, i.e. ocean currents
2 *Vertical*, i.e. the rising of bottom water and the sinking of surface water.

These movements result from the combined action of:

1 *Density* (particularly important in vertical movements)
2 *Winds* (particularly important in horizontal movements).

The density of sea water depends upon the temperature and the amount of salt in the water. It falls when water is heated, as in the tropics; when there is a large inflow of fresh water, as in the Baltic (brought by rivers), and in the polar seas (from melting ice). It rises when evaporation is high and rainfall is low, as in the Red Sea; when water is cooled, as in the polar seas, and when water freezes (salts remain in the sub-surface water which does not freeze) as in the polar seas.

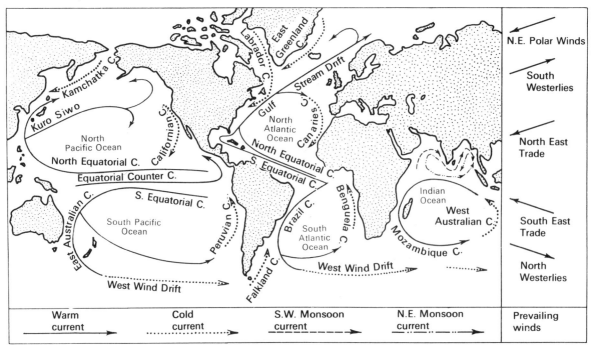

Warm current →	Cold current ·····▸	S.W. Monsoon current ———	N.E. Monsoon current ·····▸	Prevailing winds

MOVEMENTS OF OCEAN WATERS

1 Tides

Both the sun and the moon, (the latter to a greater extent), exert a gravitational attraction on the earth's surface which causes a rising and falling motion to develop in the waters of the larger oceans. This rising and falling of the water surface produces a *tide*.

The Influence of the Moon

1 Water at H_2 is 'pulled' towards the moon more than the earth – therefore water piles up at H_2 forming a high tide.
2 The earth is 'pulled' towards the moon more than the water at H_1 – therefore water lags behind and piles up at H_1 forming a high tide.
3 The moon's 'pull' causes water to be drawn from L_1 and L_2 – therefore there are low tides there.

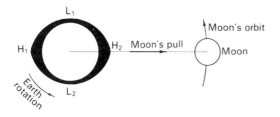

4 The rotation of the earth results in every meridian coming into the positions of two high tides and two low tides very nearly every 24 hours. The moon travels in its orbit in the same direction

as the earth is rotating and in consequence it takes about 24 hours 52 minutes, or one lunar day, for the sequence of two high and two low tides to be completed.

When the sun, earth and moon are in a straight line, as they are at Full Moon and New Moon, the gravitational force is at its greatest because the sun and the moon are "pulling" together. At these times, high tides are very high and low tides are very low, and they are called *spring tides*.

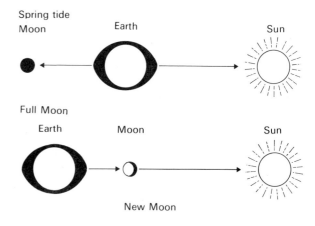

When the sun, earth and moon are not in a straight line, the sun and moon are not "pulling" together and the gravitational force is less. At Half Moon, that is, when the sun and the moon are "pulling"

at right angles, the force is at its least and the difference between high and low tides is not large. These tides are called *neap tides*.

Neap tide
Half Moon

Sun

Earth

The difference between high tide and low tide is called the *tidal range* or *amplitude*. At spring tides this is high; at neap tides it is low.

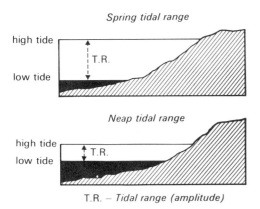

Spring tidal range

high tide

T.R.

low tide

Neap tidal range

high tide

T.R.

low tide

T.R. – Tidal range (amplitude)

Tides and Waves

At one time it was believed that the tide-producing forces resulted in the formation of two large tidal waves (H_1 and H_2 in the diagram on the opposite page) which progressively moved westward across the Southern Ocean. From these, minor tidal waves were sent out into the Pacific, Indian and Atlantic Oceans and their neighbouring seas. The two tidal waves were separated by low water (L_1 and L_2). Observations on tidal waves in different parts of the world show that this explanation can no longer be accepted. The present theory maintains that the oceans can be divided into zones in each of which the tide-producing forces cause the surface of the water to oscillate. This means that the water will rock bodily, rising and falling around the edges of the water body whilst near the centre there will be practically no rise or fall in water level (diagram below).

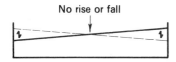

No rise or fall

Bores

When a tidal wave enters an estuary the wave increases in height as the estuary becomes increasingly shallow and narrow. Ultimately, the wave breaks and forms a wall of foaming water which often surges forward at several kilometres per hour. This usually happens when the tidal wave meets a river. Bores occur on these rivers: Hooghly, Tsien-tang-kiang (N. China) and Amazon. Bores occur in rivers which have large funnel-shaped estuaries where there is a large tidal range, and which face the direction of tidal surge.

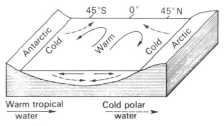

45°S   0°   45°N

Antarctic Cold   Warm   Cold Arctic

Warm tropical water   Cold polar water

2. Ocean Currents

The diagram above shows a general drift of tropical water towards the poles and a return flow of polar water towards the equator. The polar flow begins as a surface current but slowly it sinks to the ocean bottom. The winds greatly modify this simple pattern: indeed, ocean currents closely resemble prevailing winds in their direction and position. Earth rotation and the shapes of the continents also influence the direction of currents. Study the diagram below which shows the general pattern of currents for the Atlantic Ocean. Notice how water is piled up along the Brazilian Coast, and how off-shore

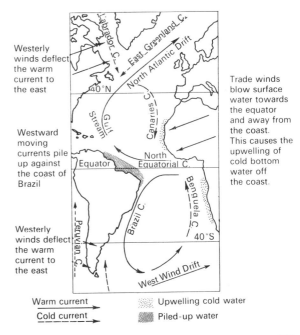

Westerly winds deflect the warm current to the east

Westward moving currents pile up against the coast of Brazil

Westerly winds deflect the warm current to the east

Trade winds blow surface water towards the equator and away from the coast. This causes the upwelling of cold bottom water off the coast.

Labrador C.

East Greenland C.

North Atlantic Drift

40°N

Gulf Stream

Canaries C.

Equator

North Equatorial C.

Peruvian C.

Brazil C.

Benguela C.

40°S

West Wind Drift

Warm current

Cold current

Upwelling cold water

Piled-up water

trade winds cause cold water to upwell along the west coast of Africa. This water forms the Canaries and Benguela Currents. The Californian and Humboldt Currents along the west coast of the Americas are formed in a similar way.

The diagram on the right indicates in greater detail how this type of cool current develops. These four currents have a cooling effect on the climates of coastal regions. Warm, moist air moving towards the coast is cooled as it passes over the cool currents and some of its moisture is condensed. Banks of mist develop along the coast. The air enters the coastal regions but there is no rain because the air is now drier. Also it is hot over the land and the air heats up and absorbs, rather than gives out moisture. Coastal mists are very common along the coast of Chile.

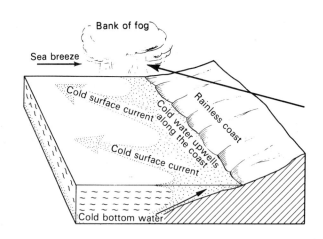

The Importance of the Oceans to Man

1 Oceans permit countries and regions to trade with one another. Goods can be moved in very large quantities by ships more cheaply than by any other means.
2 Some land margins would have colder winters if there were no warm currents in the nearby oceans. This would have an adverse effect upon such activities as agriculture.
3 The oceans contain a valuable source of food.

The map on page 105 shows the chief fishing grounds of the world. Notice that these occur in the continental shelf regions where cold and warm currents meet. Many fish feed on plankton. These plants require nitrates and phosphates, sunlight and well aerated water. Cold currents are rich in nitrates and phosphates, and epi-continental seas permit sunlight to reach almost to the bottom. The meeting of cold and warm currents causes the water to become well aerated. It is understandable therefore, why the areas shaded black form the richest fishing grounds in the world.

Ocean Currents can also Influence Climate

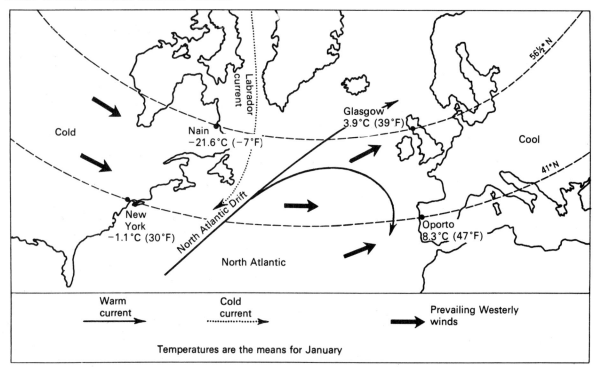

Fishing Grounds of the World

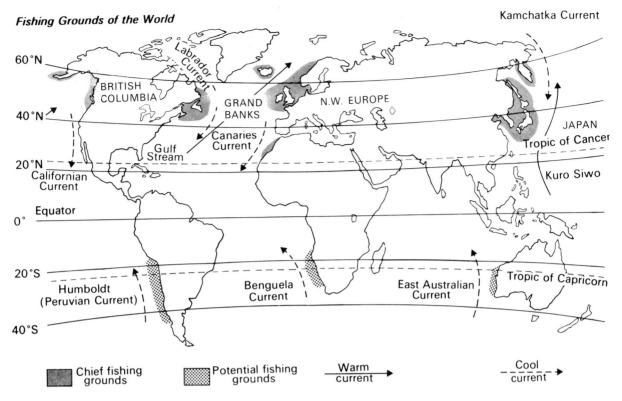

Kamchatka Current

60°N
BRITISH COLUMBIA
Labrador Current
GRAND BANKS
N.W. EUROPE
40°N
JAPAN
Tropic of Cancer
Canaries Current
Gulf Stream
20°N
Kuro Siwo
Californian Current
Equator
0°
20°S
Humboldt (Peruvian Current)
Benguela Current
East Australian Current
Tropic of Capricorn
40°S

Chief fishing grounds   **Potential fishing grounds**   Warm current   Cool current

EXERCISES

1 On an outline map of the world:
  (i) mark, name and show the direction of *two* warm and *two* cold ocean currents in each hemisphere
  (ii) mark and name *two* ocean currents whose directions are seasonally reversed
  (iii) shade *two* areas which are usually foggy for most of the year
  (iv) shade *two* coasts whose winter temperatures are raised by the combined influence of ocean currents and prevailing winds.

2 Briefly describe the relief of the ocean floor and try and account for its more important features.

3 Choose any *one* ocean current and for it:
  (i) draw a sketch-map to show its location
  (ii) briefly explain how it originates
  (iii) name any *one* other ocean current which is similar in location and origin.

4 With the aid of sketch-maps and diagrams write a brief account on *three* of the following:
  (i) the Benguela Current
  (ii) the high degree of salinity in the Dead Sea
  (iii) the seasonal change in direction of some ocean currents
  (iv) the continental shelf.

Objective Exercises

1 Which one of the following statements best describes the causes of the movements of sea water?
  A The movement in the sea water is both vertical and horizontal.
  B Differences in surface water temperature result in the formation of currents.
  C Differences in temperature and density in the oceans and seas, and the effect of winds on surface water together result in the movement in sea water.
  D Ocean currents are deflected by the rotation of the Earth.
  E Differences in the density of sea water and the shape of land masses cause movements in sea water.

  A B C D E
  ☐ ☐ ☐ ☐ ☐

2 Only one of the following currents has a warming influence upon the coast along which it flows. Which is it?
  A California Current
  B Benguela Current
  C Kuro Siwo Current
  D Humboldt Current
  E Labrador Current

  A B C D E
  ☐ ☐ ☐ ☐ ☐

3 The winter temperatures of insular north-western Europe are higher than the winter temperatures of eastern Europe in the same latitudinal zone because
  A it is on the western side of a continent
  B it is near the sea
  C it lies under westerly winds which blow over the Gulf Stream Drift
  D it receives only light falls of snow

  A B C D
  ☐ ☐ ☐ ☐

105

11   Lakes

A lake can be defined as a hollow in the earth's surface in which water collects. Some lakes are of great size and are called seas, e.g. Caspian, Dead and Aral Seas. Although most lakes are permanent, some contain water in the wet season only. Lakes in basins of inland drainage (which are usually semi-arid) may contain water for a few months only out of a period of several years.

CLASSIFICATION OF LAKES ACCORDING TO ORIGIN

It is probably true to say that the majority of lakes have been formed by the action of glaciers and ice sheets. All the others have been formed by river, marine and wind action, by earth movements and vulcanicity, and by man.

Many lakes contain fresh water but some contain saline water, e.g. the Dead Sea. Some lakes are temporary, e.g. playas and other hot desert lakes, while others are more permanent. However, all lakes, with the exception of deep rift valley lakes, are eliminated in a comparatively short period of geological time, by the forces of nature.

There are several ways in which lakes can be classified and although most lakes fit neatly into specific categories in a specific classification, some lakes, because of their origins, may fit into two such categories.

1 Lakes Produced by Earth Movements

Tectonic forces cause sagging and faulting in the earth's crust thus producing depressions, some of which may become the site of lakes. Tectonic lakes can be put into definite groups.

(i) **Lakes formed by crustal warping** These lakes occupy basin-like depressions and good examples are: the Caspian Sea (U.S.S.R.), Lake Victoria (Africa), Lake Eyre (Australia) and Lake Chad (Africa). Lake Titicaca (Peru/Bolivia) is the highest tectonic lake in the world.

(ii) **Lakes formed by faulting** Most of these lakes occur in rift valleys, e.g. Lake Nyasa and Tanganyika (East Africa), the Dead Sea and Lake Baikal (U.S.S.R.) and Loch Ness (Scotland). These lakes are usually long and narrow and very deep. The levels of some lakes are below sea level, e.g. the Dead Sea which is 393 metres (1285 feet) below sea level (its floor is 817 metres (2672 feet) below sea level).

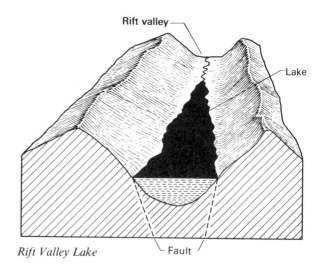

Rift Valley Lake

2 Lakes formed by Erosion

(i) **Glacially-eroded lakes**. Both valley glaciers and ice sheets can gouge out hollows and troughs on the earth's surface. These may later fill with water to form lakes. The main types of such lakes are as follows:

(a) *Cirque (corrie) lakes* Lakes often form in the armchair-like depressions, known as cirques. These lakes usually, though not always, feed mountain rivers. They are sometimes called *tarns*, and they are found in all glaciated mountain regions.

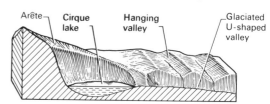

Cirque Lake   *Longitudinal section of a cirque*

(b) *Trough lakes* These occupy elongated hollows excavated in a valley bottom. They are sometimes called *ribbon lakes* because of their shape.

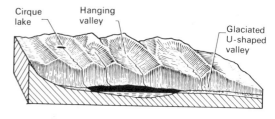

Glaciated Valley Trough Lake   *Ice-eroded trough contains a lake*

(c) *Rock basin lakes* These have been formed by ice-scouring action of ice sheets and valley glaciers which resulted in the formation of shallow hollows. They are numerous on the Baltic and Canadian Shields, and particularly good examples occur in Finland and parts of northern Canada.

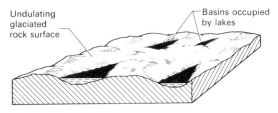

Rock Basin Lake

(ii) **Wind-eroded lakes**. Wind deflation sometimes produces extensive depressions in arid regions such as the deserts of the Southern Continents and central Asia. If the depression is excavated below the water-table then a lake will develop. The diagram shows that it is the action of eddy currents which scoop out the loose sand which is then blown away and deposited as dunes. The lakes of these depressions are not always true lakes and may be nothing more than muddy swamps. The Qattara Depression in Egypt is a good example. More permanent types of desert lakes develop when an aquifer is exposed. These are called *oases*. Some desert lakes dry up because of excessive evaporation and all that remains is a lake bed of salt. This is called a *playa* or *salt lake*.

Desert Depression Lake

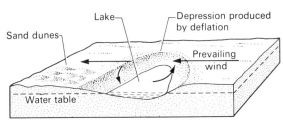

Oasis

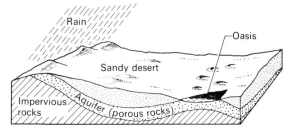

3 Lakes Produced by Deposition
River Deposits

(i) *Ox-bow lakes* Mature rivers meandering across a flood plain often produce cut-offs which in time get separated from the river to become ox-bow lakes. Such lakes are common in the lower valleys of the Mississippi, and similar rivers.

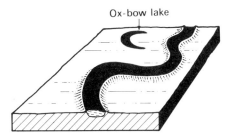

(ii) *Delta lakes* In the formation of deltas the deposition of alluvium by the rivers may cause a part of the sea to be surrounded thus turning it into a lagoon, or it may isolate a part of a distributary thus producing a lake. Lakes are common in large deltas and a good example is Étang de Vaccarès in the Rhône Delta.

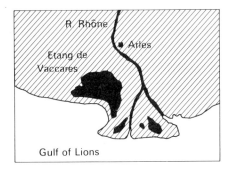

Glacial Deposits

(i) *Moraine-dammed lakes* Terminal moraines are sometimes deposited across a valley where they form a ridge which dams back the flow of water. Lakes result and these are called moraine-dammed lakes. Such lakes are common in the Lake District (England). Lake Garda (Italy) has been formed in this way.

Moraine-dammed Lake

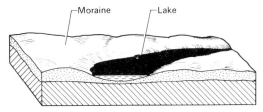

(ii) *Boulder clay lakes* Boulder clay is sometimes poorly drained and contains many depressions; lakes will often form in the latter.

Examples of such lakes occur in Northern Ireland.

Marine Deposits

Haffs These lakes have been formed by sand bars extending along a coast and cutting off indentations in the coast thus producing lagoons. Haffs are well developed on the southern coast of the Baltic Sea (see page 80). Similar lakes develop along Les Landes coast of south-west France.

4 Lakes Produced by Vulcanicity

(i) *Crater and Caldera lakes* In some violent volcanic eruptions the top of a volcano may be blown off leaving a large crater, or if this is enlarged by subsidence, a caldera. Some examples are: Crater Lake in Oregon (U.S.A.), Lake Toba in northern Sumatra and Lake Knebel (Iceland).

Crater Lake

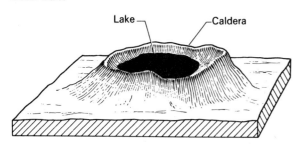

(ii) *Lava-blocked lakes* Lava flows from some volcanoes block river valleys and cause lakes to form. The Sea of Galilee was formed by a lava flow blocking the Jordon Valley.

Lava-dammed Lake

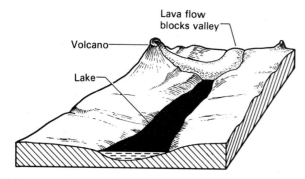

(iii) *Lava-subsidence lakes* Some lava flows are hollow and if the lava crust collapses, a depression is left which may become the site of a lake. Myvatn, in Iceland, is an example of such a lake.

5 Other Types of Lake

(i) *Solution lakes* Some rocks, especially chalk and limestone, are dissolved and removed by rain water percolating down from the surface. This action produces underground caverns in limestone regions and some of these caverns, whose floors are near to the base of the limestone layer, often contain lakes. If the cavern roof collapses the lakes become exposed. Such lakes are usually long and narrow, e.g. Lac de Chailexon in the Jura Mountains. Sometimes solution, perhaps aided by subsidence, produces large depressions called *poljes*. These often contain lakes, e.g. Lake Scutari in Yugoslavia. In Cheshire (England), rock salt has been removed by solution and shallow depressions have formed which now contain lakes, called 'flashes'.

(ii) *Barrier lakes* These can be formed by ice, lava or moraine damming a valley, as we have already seen, but damming can also be effected by landslides, avalanches or screes. Lakes formed in this way are rarely as permanent as those formed by moraine-damming.

(iii) *Beaver lakes* Some animals, especially beavers, build dams across streams and so create small lakes. Beaver Lake in Yellowstone National Park (U.S.A.) is an example.

(iv) *Mining ponds* Opencast mining results in the development of large depressions which sometimes collect water and form lakes. Tin mining in West Malaysia has created many such lakes.

(v) *Man-made lakes* By building dams across rivers, man has deliberately constructed artificial lakes, called reservoirs, e.g. Lake Mead on the Colorado River (U.S.A.)

The Value of Lakes to Man

Lakes have played an important role in the cultural and social life of peoples in many parts of the world for a very long time, and in regions where lakes are abundant, e.g. Finland and Sweden, this is especially true. The most common functions of lakes are as follows:

1 *For communication* Lakes and connecting river systems often form important natural routes for the movement of people and goods. In recent years lakes systems have been improved in some countries to enable easy movement of large ships. The Great Lakes and the St. Lawrence Waterway of North America form the longest inland water transport system in the world.

2 *For hydro-electric power development* Lakes, both natural and man-made, can, and often are used for generating hydro-electric power. Good examples are the H.E.P. plant at Jinja (Uganda) which uses water from Lake Victoria, and the

one at Grand Coulee on the Columbia River (U.S.A.) where man has created a lake which is over 80 kilometres (50 miles) long.

3 *For regulating the flow of rivers* Rivers which contain large lakes seldom flood because the lakes can absorb the run-off when rainfall is heavy, while in times of drought the water in the lakes helps to maintain a fairly steady flow of water. The Poyang and Tung Ting Lakes regulate the water-flow of the Yangtze-kiang while the Tonlé Sap regulates the water-flow of the Mekong. Many large rivers do not contain lakes but any liability of these to flood can be prevented by constructing lakes. This has been done on the Tennessee River (U.S.A.) and on the Sutlej (India), etc.

4 *As a source for fish* Some lakes are rich in fish, e.g. the Caspian Sea (U.S.S.R.) for sturgeon, the Tonlé Sap (Cambodia) and the Great Lakes (N. America), and in China and Japan man has built lakes for breeding fish.

5 *For water storage* Most urban settlements obtain their water supplies from lakes both natural and man-made. Examples of such lakes can be found in almost every country.

6 *For providing irrigation water* Some of the lakes whose water is used for generating hydro-electric power, are also used for providing irrigation water. These lakes therefore are multi-purpose. The Burrinjuck Dam on the Murrumbidgee (Australia), the Sennar Dam on the Blue Nile (Sudan) and the dams on the Tennessee (U.S.A.) all create multi-purpose lakes of this type.

7 *For developing a tourist industry* A large number of countries have used their lakes for this purpose. Switzerland, Finland, and Canada are just three examples.

8 *As moderating influences on climate* Large lakes in temperate latitudes have a moderating influence on the climate of nearby regions. In winter these lakes exert a warming influence by releasing heat which was stored up in the summer. In summer they exert a cooling influence by absorbing part of the heat. The eastern shores of Lakes Ontario, Huron and Erie have milder winters than the western shores because the winds, which blow from the west, are warmed by the lakes. Lakes also supply water vapour to winds passing over them and thus the rainfall pattern of nearby regions can be affected.

EXERCISES

1 Choose *three* of the following types of lake: ox-bow lake, crater lake, corrie lake and oasis, and for *each*:
   (i) state how it has been formed
   (ii) draw a diagram to show its main characteristics
   (iii) name *one* example or a region where an example may be found.

2 Choose *three* of the following lakes: Lake Victoria, Lake Toba, Lake Manzala, Lake Garda and the Sea of Galilee, and for *each*:
   (i) state its location
   (ii) explain its origin by means of well-labelled diagrams
   (iii) name *one* region where a lake of similar origin may be seen.

3 Choose *four* of the following features: haff, delta lake, basin of inland drainage, karst lake, tectonic lake, and a playa, and for *each* chosen feature:
   (i) briefly explain how it has originated
   (ii) draw a well-labelled diagram to show its characteristics
   (iii) name *one* example or name a region where an example may be seen.

4 A lake may be created for one or more of the following reasons:
   (i) to generate hydro-electric power
   (ii) for irrigation
   (iii) to control flooding
   (iv) to supply water to settlements.
   Choose *two* of these and for *each*:
   (a) briefly explain how it operates
   (b) name *one* specific example.

Objective Exercises

1 Which of these lakes occupies a rift valley?
   A Lake Toba
   B Lake Superior
   C Lake Eyre
   D Lake Tanganyika
   E Crater Lake

   A B C D E
   ☐ ☐ ☐ ☐ ☐

2 The levels of some lakes are below sea level. Such lakes most probably owe their origin to
   A the process of faulting
   B the damming of a glaciated valley by moraine
   C the formation of a caldera
   D the growth of a spit across a river's mouth
   E a rise in the level of the land relative to sea level

   A B C D E
   ☐ ☐ ☐ ☐ ☐

3 Which one of the following lakes owes its origin mainly to volcanic activity?
   A Lake Chad
   B Lake Baikal
   C Lake Huron
   D Lake Toba
   E Lake Superior

   A B C D E
   ☐ ☐ ☐ ☐ ☐

4 Lake Toba, in Sumatra, is
  A in a structural depression
  B in a moraine-dammed valley
  C in a depression in a limestone plateau
  D a caldera lake
  E a man-made lake

  A B C D E
  □ □ □ □ □

5 In which one of the following types of physical land-
  scape would lakes **not** develop through the process of
  silting?
  A on a deltaic plain
  B on a limestone plateau
  C on a flood plain
  D in a coniferous forest
  E on a coastal plain

  A B C D E
  □ □ □ □ □

6 Some large lakes in temperate latitudes have a mode-
  rating influence upon the climate of adjacent land sur-
  faces. Such an influence may develop when
  A the lake is high above sea level
  B the water is shallow
  C prevailing out-blowing winds cross the lake in winter
  D the lake is a basin of inland drainage

  A B C D
  □ □ □ □

7 All of the following types of lakes owe their origin to de-
  positional factors **except**
  A ox-bow lakes
  B cirque lakes
  C delta lakes
  D haffs

  A B C D
  □ □ □ □

8 Some lakes are permanent (they last for several years)
  and some are temporary (they last for a season only).
  All of the following are permanent **except**
  A cirque lake
  B playa lake
  C caldera lake
  D ox-bow lake

  A B C D
  □ □ □ □

9 In which one of the following types of region would
  you expect there to be lakes of depositional origin?
  A an arid plateau
  B an alluvial lowland
  C an ancient shield
  D limestone plateau

  A B C D
  □ □ □ □

10 A lake will probably be saline when
  A it is connected to the sea by a river
  B its surface is high above sea level
  C rivers drain into it but not out of it
  D it has a tropical latitude

  A B C D
  □ □ □ □

11 Some lakes are seasonal, that is, they dry up for a part
  of the year. Such lakes are likely to occur
  A on a deltaic plain
  B in a caldera
  C in a basin of inland drainage
  D in a glaciated valley

  A B C D
  □ □ □ □

12   Weather

When we say it is hot, or wet, or cloudy, we are saying something about the *weather*. Weather refers to the state of the atmosphere: its temperature, pressure and humidity for a place for a short period of time. If we want to find out what the weather is like we must examine:

1 temperature   2 humidity
3 pressure   4 rainfall
5 wind direction and strength   6 cloud cover
7 sunshine

MEASURING AND RECORDING WEATHER ELEMENTS

A weather station is a place where all the elements of weather are measured and recorded. Each station has a Stevenson Screen (diagrams on the right) which contains four thermometers which are hung from a frame in the centre of the screen. They are:

1 Maximum thermometer
2 Minimum thermometer
3 Wet bulb thermometer
4 Dry bulb thermometer

The screen is so built that the shade temperature of the air can be measured. It is a wooden box whose four sides are louvered to allow free entry of air. The roof is made of double boarding to prevent the sun's heat from reaching the inside of the screen, and insulation is further improved by painting the outside white. It is placed on a stand, about 121 cm (48 in) above ground level.

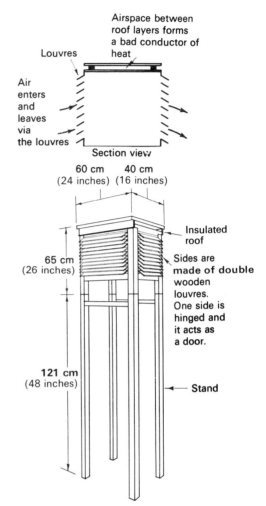

To measure maximum and minimum temperature

Maximum Thermometer

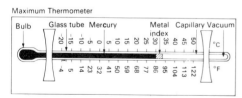

When the temperature rises the mercury expands and pushes the index along the tube.

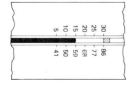

When the temperature falls the mercury contracts and the index remains behind. The maximum temperature is obtained by reading the scale at the end of the index which was in contact with the mercury. In the diagram this is 30°C (86°F). The index is then drawn back to the mercury by a magnet.

Minimum Thermometer

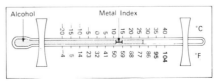

When the temperature falls the alcohol contracts and its meniscus pulls the index along the tube.

When the temperature rises the alcohol expands. The index does not move but remains in the position to which it was pulled. The minimum temperature is obtained by reading the scale at the end of the index which is nearer the meniscus. In the diagram this is 15°C (59°F). By raising the bulb of the thermometer the index is returned to the meniscus.

Six's Thermometer

This thermometer can also be used for measuring maximum and minimum temperatures. When the temperature rises the alcohol in the left-hand limb expands and pushes the mercury down this limb and up the right-hand limb. The alcohol in this limb also heats up and part of it is vaporised and occupies the space in the bulb. The maximum temperature is read from the scale on the right-hand limb. When the temperature falls the alcohol in the left-hand limb contracts and some of the alcohol vapour in the conical bulb liquefies. This causes the mercury to flow in the reverse direction. The minimum temperature is read from the scale on the left-hand limb. Note this scale is reversed.

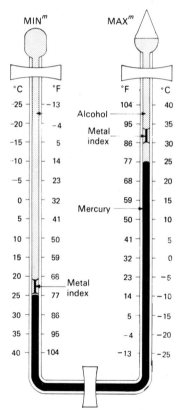

HUMIDITY OF THE AIR

No air is absolutely dry although some air, such as that over tropical deserts, contains very little water vapour. Humidity refers to the amount of water vapour in the air, but it is more important to know the relationship between the actual amount of vapour in the air and the amount of vapour the air could hold at that particular temperature. This amount is called the *Relative Humidity* (R.H.). Thus if the R.H. is 80 per cent at a temperature of 30°C (86°F), then the air is holding eight-tenths of the water vapour it could hold at that temperature. When air can hold no more vapour we say the air is *saturated* and its R.H. is 100 per cent. Now the amount of vapour air can hold is dependent upon its temperature. When this rises the air is able to hold more vapour, and when it falls it cannot hold as much. When the temperature falls it may well be that the air contains more vapour than it can hold. The excess vapour then *condenses*, i.e. turns into water droplets which form either clouds, rain, mist or fog.

Measurement of Humidity

The hygrometer shown below consists of two ordinary thermometers. The bulb of one is wrapped in a piece of muslin which dips into a container of water. This thermometer is called the wet bulb thermometer. The other is called the dry bulb thermometer. When the air is not saturated, water evaporates from the muslin and this cools the wet bulb and causes the mercury to contract. The

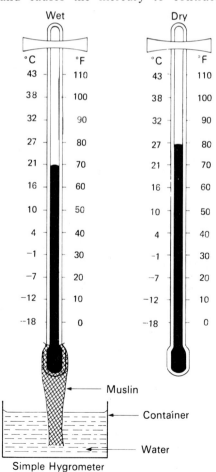

Simple Hygrometer

dry bulb thermometer is not affected in the same way, and so the two thermometers show different readings. When the air is saturated, there is no evaporation and hence no cooling. The two thermometers therefore show the same reading. The difference between the two readings is, therefore, an indication of the humidity of the air. Remember these statements:

Thermometer readings
1 No difference – air is saturated.
2 Small difference – humidity is high.
3 Large difference – humidity is low.

ATMOSPHERIC PRESSURE AND ITS MEASUREMENT

Air has weight and thus it exerts a pressure on the earth's surface. At sea level this averages 1·034 kgf/cm (14.7 lb per sq in). Pressure varies with both temperature and altitude, and the instrument which measures pressure is called a *barometer*. There are two principal types of barometer:
1 Mercury barometer; 2 Aneroid barometer

Mercury Barometer

Although this is a large and cumbersome instrument, it is very accurate and is used in many weather stations. Atmospheric pressure is read in millibars. Thus 29·92 inches of mercury are equivalent to 1013 millibars (mb) at sea level.

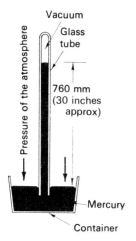

The pressure of the air on the container supports a column of mercury about 760 mm high. This amount of mercury has the same weight as a column of air about 18 kilometres high (almost 11 miles) having the same cross-sectional area as the mercury column

Aneroid Barometer

The heart of this instrument consists of a small metal box which contains very little air. The top of this box bends slightly under the influence of any change in atmospheric pressure. The movement of the box top is conveyed by a system of levers to a pointer which moves across a graduated scale. When the pressure rises the box top bends in and when the pressure falls the spring pushes the box top outwards.

Dry bulb	Wet bulb							
	20 °C	22 °C	24 °C	26 °C	28 °C	30 °C	32 °C	34 °C
20 °C	100%	–	–	–	–	–	–	–
25 °C	65%	80%	95%	–	–	–	–	–
30 °C	40%	50%	60%	80%	90%	100%	–	–
35 °C	24%	30%	35%	45%	57%	70%	82%	95%

The actual value of relative humidity is given by tables, a page of which is shown here. If the dry bulb temperature is 30°C and the wet bulb is 28°C, *i.e., depression of the wet bulb is 2°C*, the relative humidity is 90%.

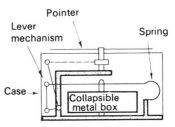

Aneroid Barometer (sectional view)

THE MEASUREMENT OF RAINFALL

All weather stations have a rain gauge. Rain falling in the funnel trickles into the jar below and at the end of a 24-hour period this is poured into a graduated measuring cylinder. The graduation of the cylinder is such that the reading obtained is the depth of rain that has fallen over an area equivalent to that of the top of the funnel. Rainfall is measured in millimetres or inches, and the cylinder is tapered at the bottom (see inset) to enable very small amounts to be measured accurately.

Position of the Rain Gauge

It must be placed in an open space so that no run-off from buildings or trees, etc., enters the funnel. It must also be sunk into the ground so that about 30 centimetres of it sticks up above ground level. This prevents rain from splashing into it from the ground, and also prevents the sun's rays from causing excessive evaporation of the water already collected in the jar.

Rain Gauge

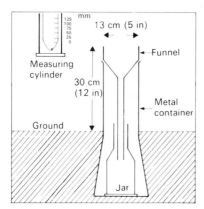

To Measure the Direction and Speed of the Wind

Wind direction is measured by a wind vane which consists of a rotating arm pivoted on a vertical shaft. The arrow of the wind vane always points in the direction from which the wind blows and the wind is named after this direction. Thus the diagram of a wind vane indicates that the wind is a north-east wind.

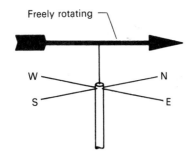

Freely rotating

W — N
S — E

Weathercock

The speed of the wind is measured by an instrument called an *anemometer*. This instrument has three or four horizontal arms pivoted on a vertical shaft. Metal cups are fixed to the ends of the arms so that when there is a wind the arms rotate. This movement operates a meter which records the speed of the wind in kilometres (or miles) per hour.

METER

ANEMOMETER

Wind velocity can also be obtained, though only approximately, by observing the way some objects are moved by the wind. This is done by referring to the *Beaufort Wind Scale*.

Beaufort Scale No.	Wind Description	Effect on land features	Speed (km/h)	Arrow Indication
0	calm	Smoke rises vertically	less than 1	
1	Light air	Direction shown by moving smoke but not by wind vane	1 - 5	
2	Light breeze	Wind felt on face; leaves rustle; wind vane moved	6 - 11	
3	Gentle breeze	Leaves and twigs in constant motion; flag moved	12 - 19	
4	Moderate breeze	Raises dust and paper; small branches moved	20 - 29	
5	Fresh breeze	Small trees begin to sway	30 - 36	
6	Strong breeze	Large branches in motion, whistling heard in telegraph wires	37 - 49	
7	Moderate gale	Whole trees in motion	50 - 60	
8	Fresh gale	Twigs broken off trees	61 - 73	
9	Strong gale	Slight structural damage occurs, especially to roofs	74 - 86	
10	Whole gale	Trees uprooted, considerable structural damage	87 - 100	
11	Storm	Widespread damage	101 - 120	
12	Hurricane	Widespread devastation experienced only in some tropical regions	above 121	

Recording of Winds
1 On a Map
Winds are shown by arrows on a weather map.

The shaft of an arrow shows wind direction, and the feathers on the shaft indicate wind force or speed. A half feather stands for a speed of 5 knots and a full feather a speed of 10 knots. A black pennant stands for a speed of 50 knots. Combinations of these can thus be used to show any speed. In practice the feathers do not have a definite value but indicate a speed between two limits. These values are given in brackets below the numbers 1, 5, and 10 and so on. Arrows are inserted at the positions of the weather stations on the map. The tip of the arrow away from the feathers points to the direction in which the wind is blowing.

II On a Wind Rose

The main purpose of a wind rose is to record wind direction for a specific place. A simple wind rose is shown in the above diagram. It consists of an octagon, each side of which represents a cardinal point. Rectangles are drawn on each side and each day when there is a wind a line is ruled across the rectangle representing the direction from which the wind is blowing. This is done for one month. The number of days when there is no wind is recorded in the circle in the centre of the octagon. The diagram shows the type of wind pattern Singapore often gets during December.

CLOUDS

When air is cooled some of its water vapour may condense into tiny droplets of water. The temperature at which the change takes place is called the *dew-point temperature*. Clouds are made of water droplets or ice particles, and so are mists and fogs which are really low-level clouds.

The shape, height and movements of clouds can indicate the type of weather which is about to occur, and because of this they are carefully studied by meteorologists who prepare weather forecasts. The following symbols are used on weather maps to indicate the amount of cloud cover. Lines drawn through places having the same amount of cloud are called *isonephs*.

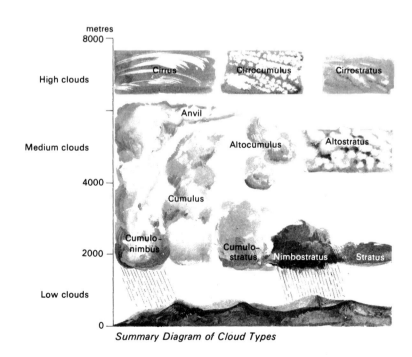

Summary Diagram of Cloud Types

Cloud Amount

Cloud Amount (oktas)	Symbol	Cloud cover description
0	○	Clear sky
1	◐	One-eighth cover
2	◑	Two-eigths cover
3	◑	Three-eighths cover
4	◑	Half of sky covered
5	◑	Five-eighths cover
6	◑	Three-quarters cover
7	◑	Seven-eighths cover
8	●	Complete cloud cover
	⊗	Sky obscured (fog)
	⊗	Missing/doubtful data

Cloud Types

Clouds can be classified according to their appearance, form and height. There are four basic groups.

1 **High Clouds**: 6000 to 12 000 metres (19 600 to 39 250 feet) above sea level
   (a) *Cirrus (Ci)*: a wispy, fibrous-looking cloud which often indicates fair weather
   (b) *Cirrocumulus (Cc)*: a thin cloud, often globular and rippled
   (c) *Cirrostratus (Cs)*: looks like a thin white sheet which causes the sun and moon to have 'haloes'

2 **Medium Clouds**: 2100 to 6000 metres (6900 to 19 600 feet) above sea level
   (a) *Altocumulus (Alt. Cu)*: globular, bumpy-looking clouds which have a flattened base; usually indicate fine weather
   (b) *Altostratus (Alt. St)*: greyish, watery-looking clouds.

3 **Low Clouds**: below 2100 metres (6900 feet)
   (a) *Stratocumulus (St. Cu.)*: low rolling, bumpy-looking clouds which have a pronounced wavy form
   (b) *Nimbostratus (Ni. St.)*: a dark, grey layered cloud which looks rainy, and which often brings rain
   (c) *Stratus (St.)*: fog-like low cloud (near to ground level); brings dull weather which is usually accompanied by drizzle

4 **Clouds of great vertical extent**: 1500 to 9000 metres (4900 to 29 500 feet)
   (a) *Cumulus (Cu.)*: a round-topped, and flat-based cloud which forms a whitish grey globular mass; indicates fair weather.
   (b) *Cumulonimbus (Cu. Ni)*: a special type of cumulus cloud which may reach up to 9000 metres (29 500 feet) and which forms white or black globular masses whose rounded tops spread out to form an anvil or cauliflower shape; thunder clouds indicating convectional rain, lightning and thunder.

Cloud types can also be shown on maps by using special symbols which are similar to those used for showing the amount of cloud cover.

Cumulonimbus Cloud

Cirrus Cloud

Cirrocumulus Cloud

Cirrostratus Cloud

Altocumulus Cloud

Nimbostratus Cloud

Cumulus Cloud

SUNSHINE

We have already discussed the various factors which affect the amount of sunshine a region receives. Latitude and position of the earth in its revolution around the sun are the more important factors.

The number of hours of sunshine a place receives can be measured by using a sensitised card, which is graduated in hours and on to which the sun's rays are focussed. Lines drawn through places having the same amount of sunshine are called *isohels*.

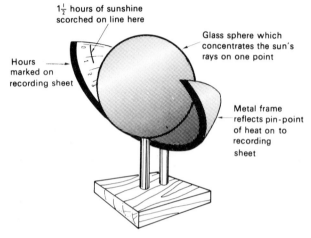

Diagram of Campbell-Stokes Sunshine Recorder

EXERCISES

1 Name the most common instruments which are contained in a weather station. Choose *two* of these, and for *each* describe: (i) how the instrument works, (ii) how its readings are taken and recorded, and (iii) the purpose of taking these readings.

2 Briefly explain the meanings of *each* of the following statements:
  (a) A Stevenson Screen should be at least 1·2 metres (or 4 feet) above the surface, its sides should be louvered and it should be placed in the open.
  (b) A Six's Thermometer is a combination of maximum and minimum thermometers.
  (c) When air is saturated wet and dry bulb thermometers should read the same.

3 Describe and name the instrument that is used for measuring rainfall. State where you would keep the instrument in order to measure rainfall accurately, and say how and when you would use it for measuring rainfall.

4 What instruments would you use to measure the following:
  (i) wind velocity
  (ii) relative humidity
  (iii) atmospheric pressure.
  Choose *two* of these instruments and for *each* draw an annotated diagram to explain how it works.

5 Study the weather data given below and then: (i) state the diurnal temperature range, (ii) state the daily mean temperature, and (iii) describe the weather conditions for the whole day giving reasons to support your answer.

Local Time	(hrs)	00.00	02.00	04.00	06.00	08.00	10.00
Temp	(°C)	12	10	10	11	12	20
	(°F)	54	50	50	51	54	68
Rel Humidity	(%)	55	58	60	59	52	49
Rainfall	(mm)	–	–	–	–	–	–
	(in)	–	–	–	–	–	–

Local Time	(hrs)	12.00	14.00	16.00	18.00	20.00	22.00
Temp	(°C)	22	25	25	21	20	16
	(°F)	72	78	78	70	68	60
Rel Humidity	(%)	44	38	32	34	39	42
Rainfall	(mm)	–	–	–	–	7	5
	(in)	–	–	–	–	0·3	0·2

Objective Exercises

1 Which one of the following is **not** an element of weather?
  A sunshine
  B cloud cover
  C height above sea level
  D rainfall
  E fog

  A B C D E
  □ □ □ □ □

2 Minimum and maximum temperatures are obtained from an instrument called
  A a barometer
  B a Six's thermometer
  C an anemometer
  D a clinical thermometer
  E a hygrometer

  A B C D E
  □ □ □ □ □

3 The relative humidity of a region is low when
  A the wet and dry bulb thermometers read the same
  B the difference between the readings of the wet and dry bulb thermometers is large
  C the temperatures are high
  D the temperatures are low
  E the wet and dry bulb thermometers read differently

  A B C D E
  □ □ □ □ □

4 Weather elements can be measured by instruments. Which one of the following pairs is **incorrect**?
  A Maximum and minimum temperatures-Six's Thermometer
  B Atmospheric pressure-Barometer
  C Wind direction-Wind Vane
  D Humidity-Rain Guage
  E Wind speed-Anemometer

  A B C D E
  □ □ □ □ □

5 Which of the following is **not** a form of precipitation?
  A snow
  B haze
  C rain
  D dew
  E hail

  A B C D E
  □ □ □ □ □

6 A Stevenson Screen usually contains all of the following **except**
  A maximum thermometer
  B wet and dry bulb thermometers
  C minimum thermometer
  D rain gauge

  A B C D
  □ □ □ □

7 In what order do the processes of saturation, evaporation and condensation take place during the formation of clouds?
  A evaporation, condensation, saturation
  B condensation, saturation, evaporation
  C saturation, condensation, evaporation
  D evaporation, saturation, condensation

  A B C D
  □ □ □ □

8 Which one of the following types of cloud rarely forms at heights lower than 5000 metres?
  A stratus
  B cumulus
  C cirrus
  D nimbostratus

  A B C D
  □ □ □ □

9 Water vapour is turned into water droplets by the process of
  A evaporation
  B liquefaction
  C convection
  D condensation

  A B C D
  □ □ □ □

10 Under which of the following conditions would the influence of aspect on temperature be most noticeable?
  A a flat sandy surface in the Sahara Desert during July
  B hilly country in the Amazon Basin in December
  C the south-facing side of a hill in Central France in April
  D the north-facing side of a hill on the Equator in June

  A B C D
  □ □ □ □

13   Climate

When we say that Malaysia is hot and wet all the year, or that Central Chile has hot dry summers and warm mild winters, we are in fact saying something about the state of the atmosphere over a long period of time. What we are really describing is the average state of the atmosphere, and such descriptions refer to the state of the *climate*. Before we make a study of the main types of climate, we must examine temperature, pressure and rainfall, etc., and find out what it is that causes these to vary from region to region. We must also know how to find the average state of these elements.

TEMPERATURE

Insolation and how Air is Heated
The sun's energy is called *insolation* and this is turned into heat at the earth's surface. Only about 45 per cent of the incoming insolation reaches the surface. The heat generated at the surface warms the air by *radiation* (heat waves sent out by the earth's surface), *conduction* (passing of heat by contact), and *convection* (passing of heat by air currents).

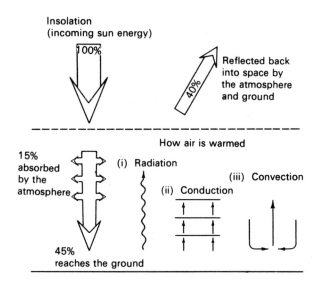

Heating and Cooling of Land and Water Surfaces
It takes over five times as much heat to raise the temperature by 2°C for a given volume of water as it does for the same volume of land. This is what the common statement 'land heats up more quickly than water' implies. The reverse of this. i.e. 'land cools more quickly than water', is equally true. In N.W. Europe, this condition causes the coastal regions to be warmed by the seas in winter and cooled by them in summer.

Water is fairly transparent and hence the sun's rays penetrate to considerable depths. Therefore, the heat which develops is more widely distributed than it is on land. Also, water movements in the seas cause the heat to be further distributed. All this means that a much greater volume of sea is heated than is the case with land.

Factors Influencing Temperature
The temperature of a place is dependent upon some or all of these factors:
1 Latitude
2 Altitude
3 Ocean currents
4 Distance from the sea
5 Winds
6 Aspect
7 Cloud cover
8 Length of day
9 Amount of dust and other impurities in the air

Latitude
The altitude of the mid-day sun is always high in the tropics and hence temperatures are always high. Outside the tropics the altitude is lower and temperatures are correspondingly lower. In general temperatures decrease from the equator to the poles. The diagram below explains this. Bands marked X contain equal amounts of sun energy, but, because area B is smaller than area A, the temperature at B will be higher than that at A. Notice also that the sun's rays at A have passed through a greater thickness of atmosphere than have those at B, and hence more sun energy arrives at B than at A.

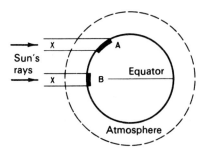

Altitude
We have already seen that the sun's rays heat the earth's surface which then passes on its heat to the air. Water vapour and dust in the air prevent this heat from rapidly escaping back into space, but at high altitudes, e.g. on the top of a high mountain, air is rarefied and it contains very little vapour or dust. The heat from the earth's surface therefore rapidly escapes and the air remains cold (see

diagram). In tropical arid regions such as hot deserts the almost complete absence of water vapour results in the earth's surface becoming intensely hot in the day. During the night most of this heat rapidly passes back into space with the result that night temperatures drop appreciably. Such regions have a large *diurnal* (daily) range of temperature.

In general, temperature falls by 6·5°C for every 1000 metres ascent, that is, 1°F for every 300 feet. A mountain which is 3000 metres (9800 feet approx.) high will have a temperature of 10·5°C (about 51°F) at its top if the temperature at its bottom is 30°C (86°F). See the diagram.

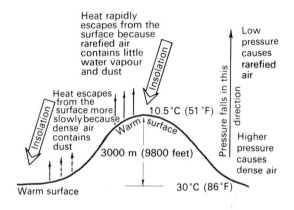

Ocean Currents

Warm and cold currents often raise or lower the temperatures of land surfaces if the winds are on-shore.

(i) Warm currents moving polewards carry tropical warmth into the high latitudes, and this warming influence is very marked in latitudes 40° to 65° on the west sides of continents, especially along the seaboard of Western Europe (page 102). The warmth is conveyed to the land by the prevailing Westerly Winds. This action is almost entirely confined to the winter season. Between latitudes 0° and 40° on the eastern sides of continents warm currents raise coastal temperatures.

Note In tropical latitudes on-shore winds crossing warm currents do not raise the temperature of the air over the land because this is already high.

(ii) Cold currents have less effect upon temperatures because they usually lie under off-shore winds (page 104). There are exceptions, e.g. the coast of Labrador, when summer temperatures are lowered by on-shore winds which blow over the cold Labrador Current.

Distance from the Sea

The sun's heat is absorbed and released more slowly by water than by land. This becomes very noticeable in temperate latitudes in the winter season when sea air is much warmer than land air. Hence on-shore winds bring warmth to coastal regions. This warming influence is confined to a narrow coastal belt because the sea air rapidly loses its heat to the colder land. Air temperatures decrease from the coast inland (middle diagram). In the summer season land surfaces are warmer than sea surfaces and the air over the land is therefore warmer than that over the sea. Therefore coastal regions are cooler than inland regions (bottom diagram). Climates whose temperatures are influenced greatly by the sea are called *maritime*, or *oceanic*, or *insular*

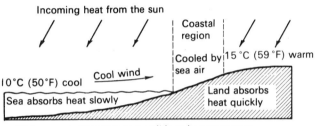

Mid-summer temperature conditions in a temperate latitude (temperatures are approximate)

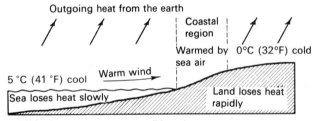

Mid-winter temperature conditions in a temperate latitude (temperatures are approximate)

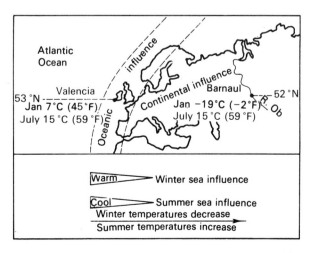

climates. These occur in coastal regions which lie under prevailing on-shore winds. Climates whose temperatures are greatly influenced by remoteness from the sea are called *continental* climates. These occur in the hearts of temperate continents.

Winds

In temperate latitudes prevailing winds from the land lower the winter temperatures but raise the summer temperature. Prevailing winds from the sea raise the winter temperatures but lower the summer temperatures. In tropical latitudes on-shore winds modify the temperatures of coastal regions because they have blown over cooler ocean surfaces. *Local winds* (see page 137) sometimes produce rapid upward or downward temperature changes.

Cloud Cover and Humidity

Clouds reduce the amount of solar radiation reaching the earth's surface and the amount of earth radiation leaving the earth's surface. When there are no clouds both types of radiation are at a maximum. The heavy cloud cover of the equatorial regions explains why the day temperatures rarely exceed 30°C (86°F) and why the night temperatures are not much lower. In hot deserts the absence of clouds and the presence of dry air result in very high day temperatures of over 38°C (about 100°F) and much lower night temperatures of 21°C (about

70°F) or below. Very humid air absorbs heat during the day and retains it at night. It also helps to prevent the loss of heat from the lower layers of the air. Thus in the humid tropics the air remains warm at night even on·days when there is little or no cloud.

Hot Desert Regions

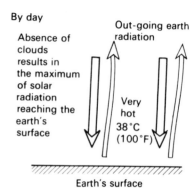

Equatorial Regions

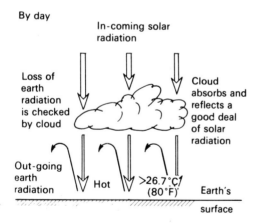

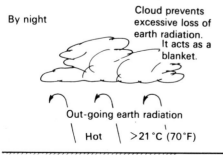

Aspect

The influence of aspect on temperature is only noticeable in temperate latitudes. In the tropics, the mid-day sun is always high in the sky and aspect is of little significance. South-facing slopes are warmer than north-facing slopes in the Northern Hemisphere, whilst in the Southern Hemisphere the reverse is true (see diagrams below).

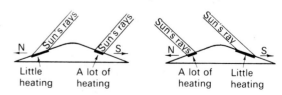

In the high latitudes, the mid-day Sun is at a low angle in the winter. Blocks of flats are usually built far apart to enable all the flats to receive some

sunshine, but in lower latitudes, flats can be built close together because the angle of the mid-day Sun is high.

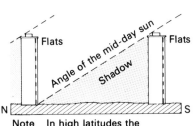

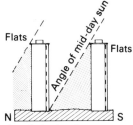

Note In high latitudes the aspect is very important in making full use of sunlight in housing developments

In low latitudes dense urban development presents less problems of light but aspect is still important in housing developments

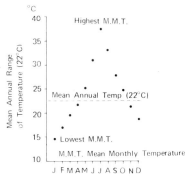

El Golea – Sahara Desert

Temperature Readings and Recording

$$\left(\frac{\text{Daily Max. Temp.} + \text{Daily Min. Temp.}}{2}\right) \text{ gives}$$

MEAN DAILY TEMPERATURE. See the diagram below.

Mean Daily Temperature is 25 C $\left(\dfrac{30° + 20°}{2}\right)$

2 (Daily Max. Temp. – Daily Min. Temp.) gives , DAILY RANGE OF TEMPERATURE. See the diagram below. Daily Range of Temperature is 10°C (30 – 20)

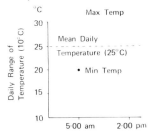

3 $\left(\dfrac{\text{Sum of Mean Daily Temps. for 1 month}}{\text{Number of days in month}}\right)$ gives MEAN MONTHLY TEMPERATURE.

4 $\left(\dfrac{\text{Sum of Mean Monthly Temps. for 1 year}}{12}\right)$ gives MEAN ANNUAL TEMPERATURE. See the diagram on the top right of this page. Annual Temperature is 22°C.

5 (Highest Mean Monthly Temp. – Lowest Mean Monthly Temp.) gives MEAN ANNUAL RANGE OF TEMPERATURE. See the diagram on the top right of this page. Mean Annual Range of Temperature is 22°C (37° – 15°).

Temperature Graphs and Maps

A Graph is used to show the temperature for a place. If the points in the diagram on the top right of this page were joined by a smooth curve, this would be a temperature graph.

A Map is used to show the temperatures for a region. The positions of all weather stations must first be plotted by dots on the map. The temperature for each station is usually adjusted to what it would be if that station were at sea level. This is done by adding 6·5°C for every 1000 metres of height for the station. The adjusted temperature values are then written alongside the dots and all places having the same temperature are joined by a smooth line. Such a line is called an *isotherm*. Isotherms rarely pass through a station and they must be inserted by using interpolation which is based on proportion.

25° isotherm will pass midway between points 24° and 26°

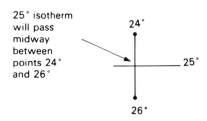

Isotherms in degrees C

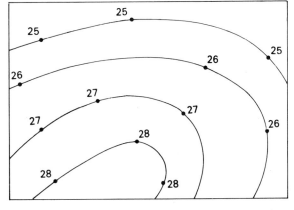

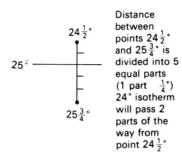

Distance between points $24\frac{1}{2}°$ and $25\frac{3}{4}°$ is divided into 5 equal parts (1 part $\frac{1}{4}°$) 24° isotherm will pass 2 parts of the way from point $24\frac{1}{2}°$

A

Temperature °C	State of the Air
Below − 10°	Very Cold
− 10° to 0°	Cold
0° to 10°	Cool
10° to 21°	Warm
21° to 30°	Hot
Over 30°	Very Hot

B

Annual Range of Temperature °C	Description
Below 3°	Negligible
3° to 8°	Small
8° to 19°	Moderate
19° to 30°	Large
Over 30°	Very Large

How to Describe the Temperature State of the Air

When we talk about the climate of a region we nearly always use such adjectives as very hot, hot, warm or cool, etc. It is important therefore to give some temperature value to each of these adjectives. In the table A on page 111 the temperature refers to daily, monthly or annual mean temperature. According to this table a month which has a mean temperature of 25°C will be referred to as a hot month. A similar table B can be used for describing the mean annual range of temperature.

World Distribution of Temperature

The following two maps show the mean temperatures for the months of January and July respectively. You should examine the isotherms of these two maps very carefully and then try to account for the following statements which can be made about these two maps.

July Temperatures in °C (°F)

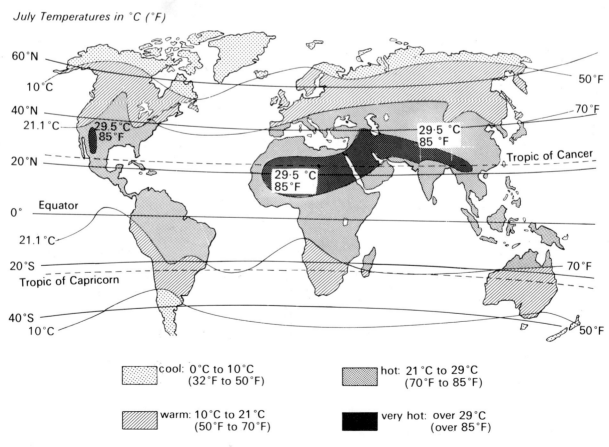

cool: 0°C to 10°C (32°F to 50°F)

warm: 10°C to 21°C (50°F to 70°F)

hot: 21°C to 29°C (70°F to 85°F)

very hot: over 29°C (over 85°F)

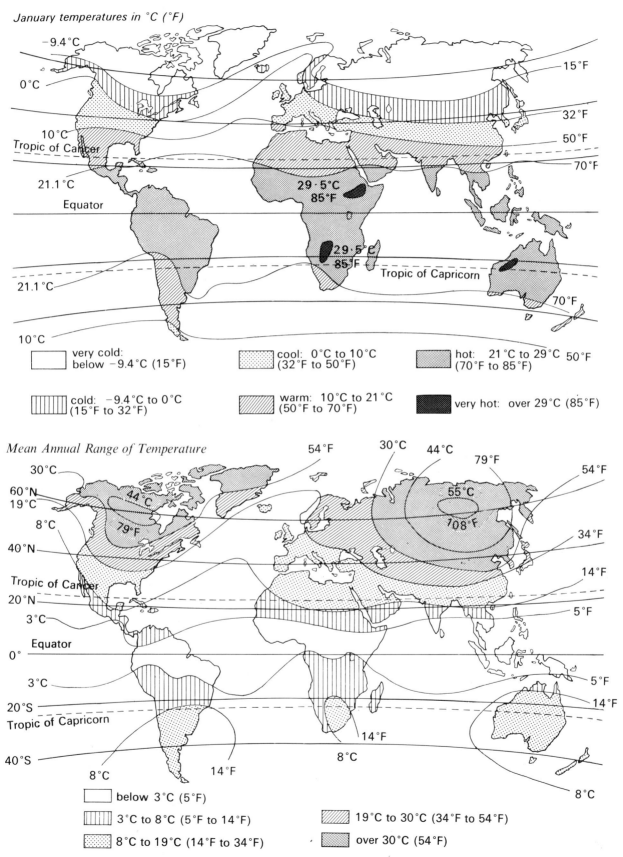

January temperatures in °C (°F)

−9.4°C
15°F
0°C
32°F
10°C
50°F
Tropic of Cancer
70°F
21.1°C
29·5°C
85°F
Equator
29·5°C
85°F
Tropic of Capricorn
21.1°C
70°F
10°C

very cold:
below −9.4°C (15°F)

cool: 0°C to 10°C
(32°F to 50°F)

hot: 21°C to 29°C 50°F
(70°F to 85°F)

cold: −9.4°C to 0°C
(15°F to 32°F)

warm: 10°C to 21°C
(50°F to 70°F)

very hot: over 29°C (85°F)

Mean Annual Range of Temperature

54°F
30°C
44°C
79°F
30°C
54°F
60°N
19°C
44°C
55°C
8°C
79°F
108°F
34°F
40°N
14°F
Tropic of Cancer
20°N
5°F
3°C
Equator
0°
3°C
5°F
20°S
14°F
Tropic of Capricorn
14°F
40°S
8°C
8°C
14°F
8°C

below 3°C (5°F)

3°C to 8°C (5°F to 14°F)

19°C to 30°C (34°F to 54°F)

8°C to 19°C (14°F to 34°F)

over 30°C (54°F)

125

1 There is a definite northward movement of all isotherms between January and July.
2 This movement of the isotherms is greater over the land than it is over the oceans.
3 The highest temperatures for both January and July are over the continents.
4 The lowest temperatures for January are over the northern continents (Asia and North America).
5 The isotherms bend poleward over the oceans but equatorward over the continents in January.
6 The isotherms bend equatorward over the oceans but poleward over the continents in July.
7 The seasonal changes are less marked over the southern continents than over the northern continents.

The mean monthly temperatures for January and July can be used to prepare a map to show the mean annual ranges of temperatures. This has been done in the map on page 125 (bottom). The map indicates the importance of maritime and continental influences which will be discussed later. In the meantime study the map and, as with the previous two maps, try to account for these statements.

1 The range of temperature in general increases from the equator to the poles.
2 The greatest range of temperature occurs not at the poles but over Asia and North America in latitude 60°N (approx.).
3 Coastal regions have a smaller range of temperature than do continental interiors.
4 The range of temperature on the eastern sides of Asia and North America is greater than it is on the western sides in the *same latitude*.

PRESSURE

We have already seen that air has weight and therefore it exerts a pressure, called *atmospheric pressure*, on the earth's surface. This pressure is not the same in all regions nor is it always the same in one region all the time. Atmospheric pressure depends primarily on three factors:
1 Altitude
2 Temperature
3 Earth Rotation

Influence of Altitude on Pressure

The pressure of air at ground level is higher than that of air at the top of a high mountain. The air at ground level has to support the weight of the air above it and the molecules in the bottom air must push outwards with a force equal to that exerted by the air above it. The molecules of air at the top of a mountain are pushing outwards with much less force because the weight of the air above it is less. It therefore follows that *when air*

sinks its pressure increases. In sinking, the volume of the air decreases but the number of molecules in it remains the same. The outward pressure of these molecules is spread over a smaller area. Similarly, *when air rises* its volume increases and the outward pressure of its molecules is spread over a larger area and *its pressure decreases*.

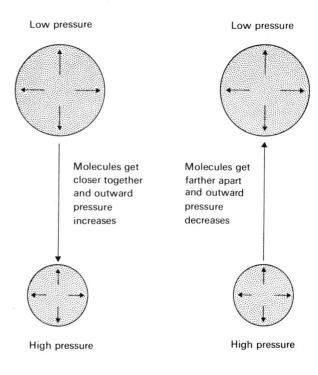

Low pressure   Low pressure

Molecules get closer together and outward pressure increases

Molecules get farther apart and outward pressure decreases

High pressure   High pressure

Very little air above this height

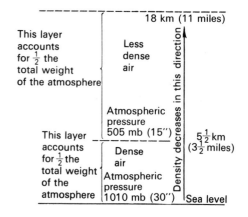

18 km (11 miles)

This layer accounts for $\frac{1}{2}$ the total weight of the atmosphere — Less dense air

Atmospheric pressure 505 mb (15")

This layer accounts for $\frac{1}{2}$ the total weight of the atmosphere — Dense air

Atmospheric pressure 1010 mb (30")

Density decreases in this direction

$5\frac{1}{2}$ km ($3\frac{1}{2}$ miles)

Sea level

Influence of Temperature on Pressure

1 We have seen that when air sinks its pressure increases because it becomes compressed. And when it becomes compressed its molecules move

more quickly and heat is produced. *The temperature of air therefore rises when its pressure rises.*

2 When air rises its pressure decreases because it expands. When air expands its molecules move more slowly and heat is used up. *The temperature of air therefore falls when its pressure falls.*

3 When air is heated it expands and when this happens the outward pressure of its molecules is spread over a larger area. This means that the pressure of the air decreases. *The pressure of air therefore falls when its temperature rises.*

4 When air is cooled it contracts and when this happens the outward pressure of its molecules is spread over a smaller area. This means that the pressure of the air increases. *The pressure of air therefore rises when its temperature falls.*

Effects of temperature on pressure

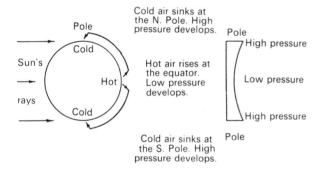

Other factors affect pressure

Supposing that only temperature affected pressure, then the pressure pattern of the atmosphere would be something like that shown in the diagram above. There would be a belt of low pressure around the earth at the equator, and two belts of high pressure, one over each pole. But because altitude, temperature and earth rotation all affect pressure, the pressure pattern is not as simple as this.

Influence of Rotation on Pressure

The rotation of the earth causes air at the poles to be thrown away towards the equator. In theory this should result in air being piled up along the equator to produce a belt of high pressure, whilst at the poles low pressure should develop as shown in the top right diagram. But what actually happens is much more complicated and we must now try and find out how temperature and rotation together affect the pressure pattern.

Effects of Earth rotation on pressure

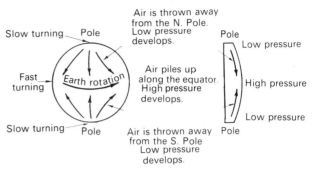

Combined Influence of Rotation and Temperature on Pressure

First, the effect of temperature

1 Low temperatures at the poles result in the contraction of air and hence the development of high pressure (see page 128)

2 High temperatures along the equator result in the expansion of air and hence the development of low pressure (see page 128). This is called the *Doldrum Low Pressure.*

Second, the effect of rotation

1 Air blowing away from the poles crosses parallels which are getting longer. It therefore spreads out to occupy greater space, that is, it expands and its pressure falls. These low pressure belts become noticeable along parallels 60°N and 60°S. As air moves away from the poles, more air moves in from higher levels to take its place. Some of this comes from the rising low pressure air in latitudes 60°N and 60°S.

2 Air rising at the equator spreads out and moves towards the poles. As it does so it crosses parallels which are getting shorter and it has to occupy less space. It contracts and its pressure rises. This happens near to latitudes 30°N and 30°S and in these latitudes the air begins to sink thus building up the high pressure belts of these latitudes. These are called the *Horse Latitude High Pressures.*

Note Some of the high pressure air in latitudes 30°N and 30°S moves over the surface towards the equator, and some of it moves towards the poles. The air moving towards the equator replaces the air which is rising there. The air moving towards the poles reaches latitudes 60°N and 60°S where it replaces the air which is rising there.

If the earth had a uniform surface, i.e. if it was all land or all water, then the pressure pattern would be as shown in the diagrams on page 128. Winds have been inserted because they are produced by

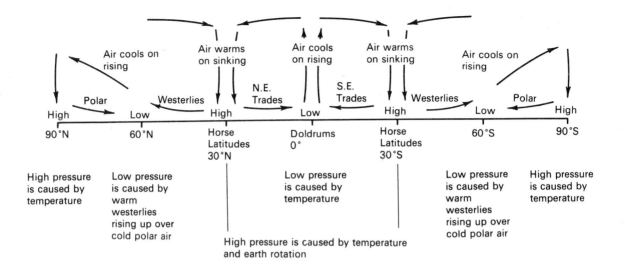

pressure systems. You can see from the diagrams that at the surface, winds blow from highs to lows, whilst at high levels they blow from lows to highs. You can also see that there are three wind systems in each hemisphere: one operates between the pole and latitude 60°, one between latitude 60° and latitude 30° and one between latitude 30° and the equator. These three systems are called the *Polar*, the *Tropical* and the *Equatorial systems*.

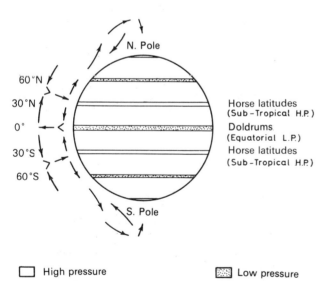

The Actual Pressure Systems on the Earth as it is

The earth's surface is not uniform, but is composed of land and water, and its axis is tilted at an angle of $66\frac{1}{2}°$. It has already been shown how land and water surfaces heat and cool at different rates, and

also how temperatures in regions outside the tropics and especially over land surfaces vary considerably from season to season. All of this results in changes in the pressure systems as given in the diagram above. Study the pressure maps on page 129 which are for January and July and try to account for the differences.

Pressure in January

(i) The Equatorial Low Pressure Belt extends well into the Southern Hemisphere where it is the summer season.

(ii) Low pressure is particularly well developed over Australia.

(iii) The low temperatures over the hearts of the northern continents produce strong high pressure systems. These link up with the sub-tropical high pressure systems.

(iv) The sub-tropical high pressures of the Southern Hemisphere are formed only over the oceans.

(v) Low pressure systems are well developed over the North Atlantic and the North Pacific Oceans.

Pressure in July

(i) The Equatorial Low Pressure Belt extends well into the Northern Hemisphere where it is the summer season, and it links up with the low pressure belts over north-west India and Pakistan and south-west U.S.A. Pressure in these areas is very low.

(ii) The sub-tropical high pressure belt in the Northern Hemisphere is no longer continuous, and it now exists as separate cells of high pressure only over the oceans.

(iii) The sub-tropical high pressure cells of the Southern Hemisphere combine to form one belt of high pressure which extends across the three continents.

(iv) The low pressure cells over the North Atlantic and North Pacific Oceans are poorly developed and have moved north.

Points to remember about the Seasonal Changes in Pressure

1 The revolution of the earth and the permanent tilt of its axis result in the overhead sun 'moving' between the Tropics as shown in the section dealing with the seasons. This causes the Doldrums to move north and south of the equator.

2 The Doldrums is the key to the pressure systems and hence these also move north and south of the positions they occupy when the sun is overhead along the equator (equinoxes).

3 Seasonal temperature changes over the continents in the Northern Hemisphere cause seasonal pressure changes over these continents.

Recording Pressure on Maps

The atmospheric pressures recorded at the weather stations in a region, at approximately the same time, are plotted on a map of the region at the positions of the stations. Lines are then drawn through places having the same pressures. These lines are called *isobars*.

Note: The pressures are usually 'reduced' to sea level before they are plotted. The pressures are expressed in millibars (mb).

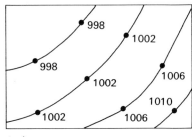

Isobars

WINDS

In our work on pressure we have studied the relationship between temperature and pressure, and we have seen that:

1 A rise in temperature causes air to expand and its density to decrease.

2 A fall in temperature causes air to contract and its density to increase.

These statements mean that *high temperatures give rise to low pressure at sea level, and low temperatures give rise to high pressure at sea level.*

Land and Sea Breezes

When a region is hotter than a neighbouring region air moves into the hot region from the cooler region to take the place of the hot air which has expanded

January Pressure

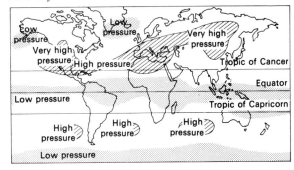

July Pressure

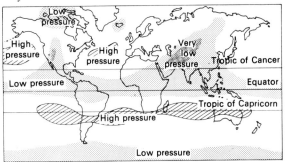

and risen. The air which moves in is a *wind*. The diagram below explains how this comes about. During the day the land gets warmer than the sea and hence air pressure is lower over the land than the sea. Air blows from sea to land as a *sea breeze*. During the night the land cools more quickly than the sea and the reverse process sets in. Land and sea breezes illustrate this process particularly well.

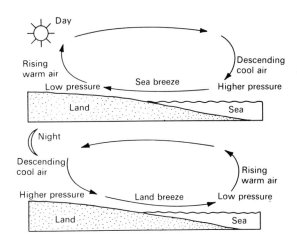

Earth Rotation Influences Wind Direction

In an earlier part of this book (page 10) we saw that

the rotation of the earth causes freely moving water and air masses to be deflected from their original courses. This is summarised by *Ferrel's Law* which states that freely moving bodies are deflected to their *right* in the Northern Hemisphere, and to their *left* in the Southern Hemisphere. In the diagram below dotted lines represent the paths which the winds would take if the earth was not rotating. Winds therefore do not follow straight paths, but curving ones as shown in this diagram.

In this diagram the earth is regarded as being uniform, i.e. all land or all water.

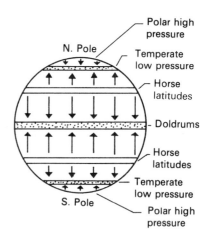

How the Winds would blow on a non-rotating earth

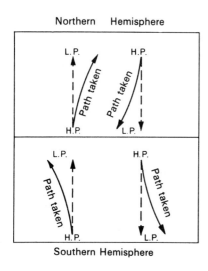

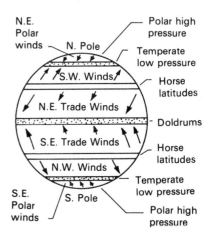

How the Winds blow on a rotating earth

Prevailing Winds

A wind which blows more frequently than any other wind in a particular region is called a *prevailing wind*. The bottom right diagram shows the prevailing winds of the world. You will notice from this diagram that winds are named by the direction from which they blow.

The maps on the top of page 131, which show the wind patterns for January and July respectively, should be carefully studied. Notice the following points:

(i) Over North America and Asia the winds are *out-blowing* during the winter when pressure is high, but are *in-blowing* during the summer when pressure is low.

(ii) The Westerly Winds in the Northern Hemisphere blow more persistently in winter than in summer. The Westerly Winds in the Southern Hemisphere blow persistently throughout the year.

(iii) The North East Trades over Asia are strengthened by the Asian High Pressure in winter, but completely disappear in summer because of the Asian Low Pressure.

(iv) The North East Trade Winds in eastern U.S.A. disappear in the summer because of the American Low Pressure.

(v) All the wind belts in July shift slightly northward of their January positions.

We have seen that because the earth's surface is part land and part water, changes take place in the pressure belts from season to season (page 129). This is accompanied by changes in the pattern of winds. The basic wind pattern as shown in the above diagram can be recognised in the wind patterns for both January and July. It only really breaks down over Asia and North America where winds change direction from season to season.

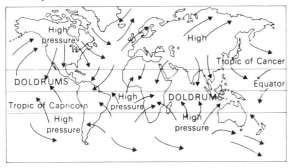

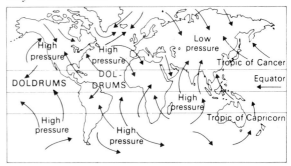

Monsoon Winds

Monsoon is derived from '*mausim*' (Arabic) which means *season*, and this word is applied to winds whose direction is reversed completely from one season to the next. Monsoon winds are best developed over Asia (Indian sub-continent, S.E. Asia, China and Japan).

In the map on the right and in that below, the full black arrows represent monsoon winds. The arrows in broken lines represent non-monsoon winds.

July

1 The Himalayas separate the Asian Low Pressure from the Punjab Low Pressure.
2 Winds blow from the **Australian High Pressure** across the Equatorial **Low Pressure to the more** intense Asian **Low Pressure.**
3 Winds blow from the Horse Latitudes High Pressure across the Equatorial Low Pressure to the more intense Punjab Low Pressure.

January

1 The Himalayas separate the Asian High Pressure from the Punjab High Pressure.
2 Winds blow from the Asian High Pressure across the Equatorial Low Pressure to the more intense Australian Low Pressure.
3 Winds blow from the Punjab High Pressure to the Equatorial Low Pressure.

January

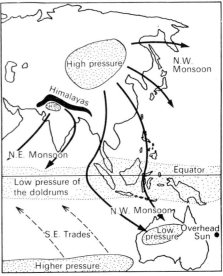

July

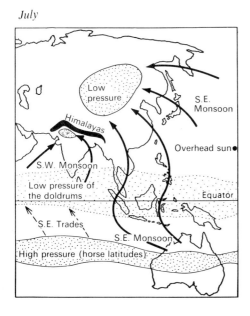

Characteristics of the Prevailing Winds

Polar Winds

1 They blow from the Polar High Pressures to the Temperate Low Pressures.
2 They are better developed in the Southern Hemisphere than in the Northern Hemisphere.
3 They are deflected to the right to become the N.E. Polar Winds in the Northern Hemisphere and to the left to become the S.E. Polar Winds

in the Southern Hemisphere.

4 They are irregular in the Northern Hemisphere.

Westerlies

1 They blow from the Horse Latitudes to the Temperate Low Pressures.

2 They are deflected to the right to become the S. Westerlies in the Northern Hemisphere and to the left to become the N. Westerlies in the Southern Hemisphere.

3 They are variable in both direction and strength.

4 They contain depressions.

Trades

1 The word 'trade' comes from the Saxon word *tredan* which means to tread or follow a regular path.

2 They blow from the Horse Latitudes to the Doldrums.

3 They are deflected to the right to become the N.E. Trades in the Northern Hemisphere and to the left to become the S.E. Trades in the Southern Hemisphere.

4 They are very constant in strength and direction.

5 They sometimes contain intense depressions.

Depressions and Anticyclones
Depression

This is a mass of air whose isobars form an oval or circular shape, where pressure is low in the centre and increases towards the outside. Depressions are well developed in the Westerly Wind and sometimes in the Trade Wind Belts. Depressions are rarely stationary and they tend to follow definite tracks. They are most influential in maritime areas because they weaken over land areas.

Anticyclone

This is a mass of air whose isobars form a similar pattern to that of a depression, but in which pressure is high at the centre, decreasing towards the outside. Anticyclones often remain stationary before gradually fading out. They often affect whole continents.

The wind circulation in depressions and anticyclones is shown in the following diagrams. You will see that *air moves anti-clockwise in a depression and clockwise in an anticyclone in the Northern Hemisphere, and clockwise in a depression and anti-clockwise in an anticyclone in the Southern Hemisphere.*

Types of Depressions

There are two types:

1 Depressions (temperate cyclones).

2 Tropical cyclones (hurricanes, typhoons and willy-willies).

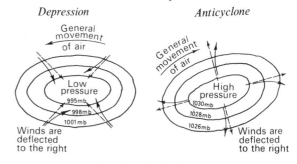

Northern Hemisphere

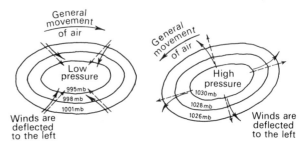

Southern Hemisphere

Depressions (temperate cyclones).

These arise in the belt of Westerly Winds and are caused by the mixing of cold air from polar regions with warm, humid air from tropical regions. They consist of swirling masses of air (anti-clockwise in N. Hemisphere and clockwise in S. Hemisphere). They usually bring prolonged rain to coastal regions and often very windy weather.

Tropical Cyclones (hurricanes and typhoons)

These arise in the belt of Trade Winds where these winds begin to disappear in the Doldrums. They move in a general westerly direction and they have very low pressure. Because of this they give rise to winds of great force which are extremely destructive. Their circulation is the same as that of the depressions. In Asia they are called *typhoons*; in the West Indies they are called *hurricanes*; off the coast of Queensland they are called *willy-willies*.

How depressions and cyclones develop

The earth's atmosphere can be divided into several air masses each of which has distinctive temperature and humidity characteristics. Two such air masses are the polar air mass which originates over unfrozen land and water in high latitudes and which moves towards tropical latitudes, and the tropical air mass which originates in the low latitudes and which moves poleward.

The zone along which two contrasting air masses meet is called a *front*. The best known fronts are the *polar front* (between polar and tropical air masses) and the *inter-tropical front* (between the trade wind belts of the Northern and Southern Hemispheres).

When contrasting air masses lie adjacent to one another, indentations develop along the front separating them either because the warmer air of one air mass slowly rides up over the colder air of the other, or because the colder air pushes into the warmer air and forces it to rise. The positions of the two main fronts, the polar front and the inter-tropical front varies from season to season.

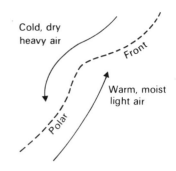

(i)

Note: Width of developing depression: 320 to 800 kilometres (200 to 500 miles)

The development of a depression

Stage 1 Along the polar front, cold polar air moves in a general westerly direction, and warm tropical air moves in a general easterly direction. The frictional effects of the two air flows causes a wave to develop. See diagram (i).

Stage 2 The wave bulges into the colder air and gets larger. Pressure falls at the tip of the wave and an anti-clockwise circulation of winds blows around this low pressure point in the Northern Hemisphere as shown in diagram (ii). The circulation is clockwise in the Southern Hemisphere.

Stage 3 As the bulge develops, the warm air rises up over the colder air at the front of the bulge. This front is called the *warm front*. At the rear of the bulge, the colder air forces its way under the warm air. The rear is called the *cold front*. The warm air between the two fronts is called the *warm sector*. See diagram (iii). You can see from diagram (iv) that the warm front is much more gently sloping than the cold front. Eventually the cold front catches up with the warm front and lifts it off the ground. It then becomes an *occluded front* and it soon dies out.

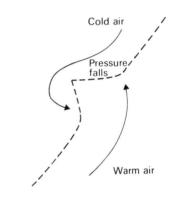

(ii)

The weather associated with a depression

In diagram (v) the depression is approaching A and the sequence of weather at A will be a follows:

1 Clear sky except for a little high cirrus cloud. The wind is from the south-east. After a while a definite cloud cover develops and light showers of rain occur, which get progressively heavier. The warm front passes.

2 The rain stops and the wind changes direction from south-east to south-west. Temperatures rise and the air is humid (warm sector).

3 As the cold front passes the weather changes very rapidly. The wind now blows from the north-west and the temperature falls. Short but often heavy falls of rain occur. With the passage of the depression the sky clears and it remains cool.

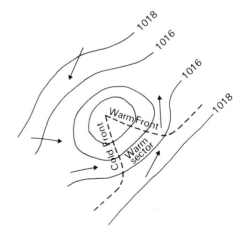

(iii)

(iv)

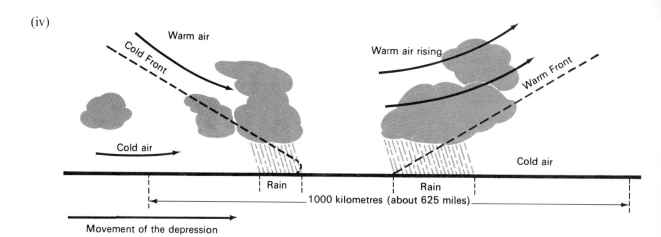

(v)

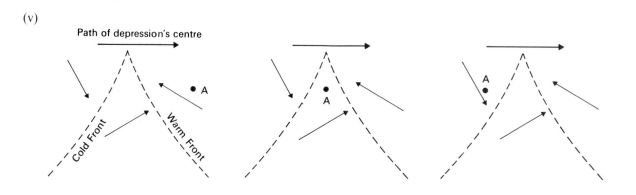

Weather Maps and Depressions

The following two maps show how warm and cold fronts appear on a weather map.

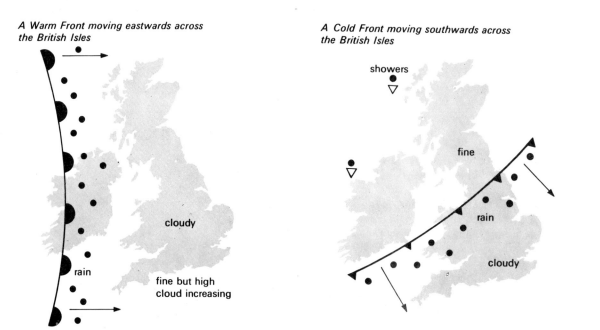

134

This diagram shows the birth of a depression. Cold air meets warm moist air moving in the opposite direction. The cold air forms a gently sloping wedge underneath the warm air and a front develops.

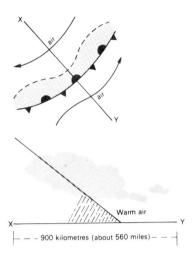

The next two diagrams show a mature depression as it appears on a weather map, and in three-dimensional form.

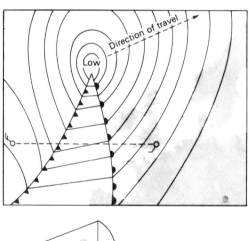

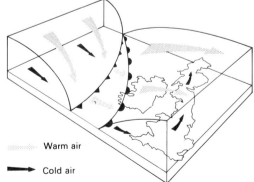

The weather map on the right shows a depression and gives information on all aspects of weather associated with it. This type of map is often called a *synoptic chart*.

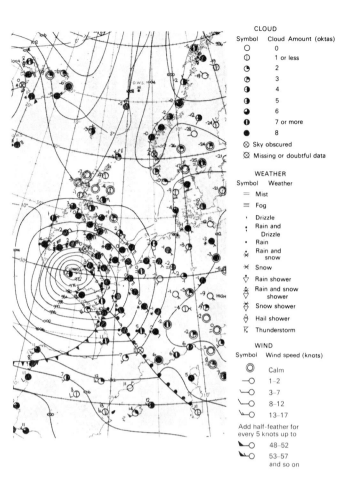

How a tropical cyclone develops

Tropical cyclones develop along the inter-tropical front where the trade wind air masses converge. These air masses over the oceans have moist lower layers but drier upper layers. When two such air masses meet, one of them will tend to be lifted up. This causes *instability*. Large cumulus clouds develop which produce intense thundery conditions. The actual origin of a tropical cyclone is not fully clear.

Location and Movement of Depressions and Tropical Cyclones

Tropical Cyclone

The diagram on page 137 shows that a tropical cyclone is funnel-shaped. The air on the outside is rapidly rising and swirling in an anti-clockwise direction in the Northern Hemisphere. The air inside the funnel is relatively calm because it is descending.

The Nature of a Tropical Cyclone

1 Before the cyclone arrives the air becomes very still, and temperature and humidity are high.

2 As the front of the vortex arrives gusty winds

135

Location and movement of cyclones

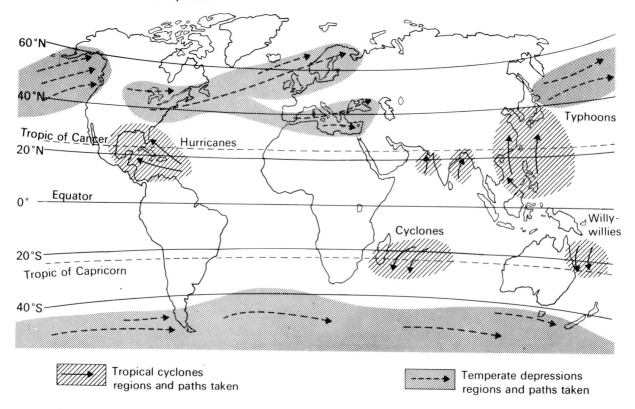

Tropical cyclones regions and paths taken

Temperate depressions regions and paths taken

develop and thick clouds appear.

3 When the vortex arrives, the winds become violent (upward surges) and often reach 250 kilometres (160 miles) per hour. Dense clouds and torrential rain reduce visibility to a few metres.

4 Calm conditions return when the eye of the cyclone arrives.

5 The arrival of the rear of the vortex brings in violent winds, dense clouds and heavy rain. The wind is now blowing from the opposite direction to that of the front of the vortex.

Where Tropical Cyclones form and what happens to them

They always form over the oceans in the tropics where the northerly and southerly Trade Winds meet. They follow an easterly course and soon lose their strength when they cross the coast and penetrate inland.

Section through a Tropical Cyclone

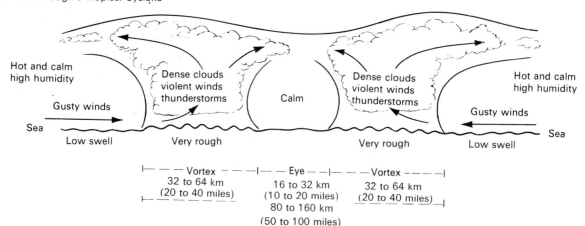

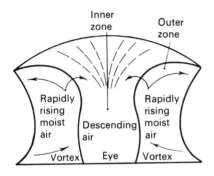

Section through a Tropical Cyclone

The different names given to tropical cyclones appear in the map on page 136.

Tornadoes, which occur in the Mississippi Valley of U.S.A., have not been shown on this map. A tornado differs from a tropical cyclone in that it forms over land. It is more destructive than a cyclone because its winds often exceed 320 kilometres (200 miles) per hour. Fortunately tornadoes are only a few hundred yards across.

Note The air moves into the depression from all directions.

Local Winds (affect only limited areas and blow for short periods of time)

Most local winds are developed by depressions. The air circulation in these is such that air is drawn in from tropical regions in the front of the depression (this gives rise to hot winds), and from polar regions in the rear of the depression (this gives rise to cold winds). See the diagram below.

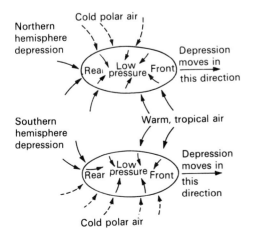

Tornado passing across southern U.S.A. The uprising funnel of air is clearly visible. Although this is only a few hundred metres wide at the most it can cause tremendous damage by its 'vacuum' effect.

137

A Depression Winds
Hot Winds

Usually these are both hot and dusty, but if they have crossed a sea surface they become very humid.

Cold Winds

Often very strong and gusty and bitterly cold.

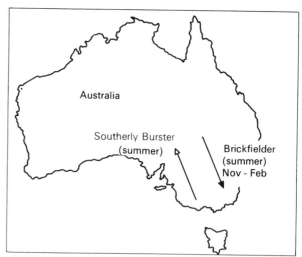

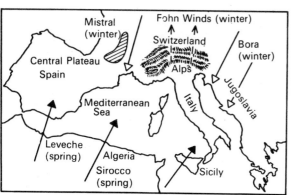

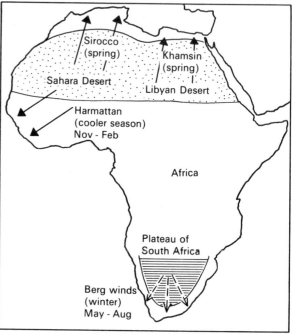

B Descending Winds

These are warm winds which descend mountain slopes onto the lowlands. As the air rises up the windward side of a mountain it cools at the rate of 6·5°C per 1000 metres. When condensation occurs, heat is given out, and the air cools at 4·5°C per 1000 metres. After crossing the mountain the air descends and it warms as it does so. Warming takes place at the rate of 6·5°C per 1000 metres. In the left diagram (see page 139) air on the windward side of the mountain at sea level is 7°C whereas on the leeward side at the same level it is 9°C.

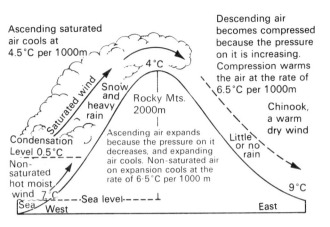

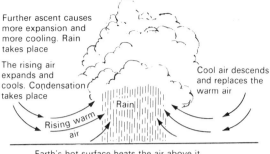

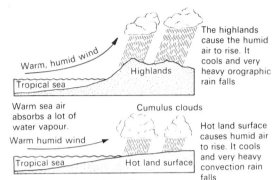

The highlands cause the humid air to rise. It cools and very heavy orographic rain falls

Warm sea air absorbs a lot of water vapour.

Warm humid wind

Tropical sea

Cumulus clouds

Hot land surface causes humid air to rise. It cools and very heavy convection rain falls

Note: Because tropical sea air is almost saturated only a little cooling is required to condense the water vapour into rain

C Convection Winds

In some hot deserts there are violent convection currents caused by intense heating. These produce convection winds which give rise to dust and sand storms. The Dust Devils of the Sahara which also affect West Africa and the Simoon of the Sahara, are caused by convection winds.

Summary of Local Winds
Depression Winds

Hot Winds	Cold Winds	Descending Winds
Sirocco	Mistral	Chinook
Leveche	Bora	Föhn
Khamsin	Pampero	Berg
Harmattan	Southerly Burster	Nor'wester
Santa Ana	Buran	Samun (Persia)
Zonda	*Convection Winds*	Nevados (Ecuador)
Brickfielder	Dust Devils	
	Simoon	

RAINFALL

Under certain conditions water vapour, which is a gas, takes the form of tiny droplets of water. These appear as a cloud, mist, fog, hail, dew or rain. All of these forms are referred to as *precipitation*.

Conditions necessary for Precipitation to Occur

1 The air must be saturated.
2 The air must contain small particles of matter such as dust around which the droplets form.
3 The air must be cooled below its dew-point.

How Air is Cooled

Air is cooled in two main ways:
1 *By being made to rise* – (most of the world's rain results from this type of cooling)
   (i) Hot air rising by convection currents (top diagram)
   (ii) A wind blowing over a mountainous region (middle diagram)
   (iii) Warm air rising over cold air.

2 *By passing over a cold surface* – (most of the world's mist and fog results from this type of cooling)
   (i) A warm wind blowing over a cold current
   (ii) A warm wind blowing over a cold land surface.

Oceans and ocean currents influence rainfall

The importance of oceans as a source of rainfall is well known. In tropical latitudes the air over the oceans contains much more water vapour than does the cooler air over the oceans in temperate latitudes. But remember not all tropical regions have heavy or even fairly heavy rainfall.
   (i) Winds blowing over a warm current on to a cooler land surface usually bring heavy rain.
   (ii) Winds blowing over a cool current on to a warmer land surface usually bring little or no rain.
   (iii) Winds blowing over a warm current and then over a cold current usually produce fog.

Types of Rainfall
I Convection Rain

Convection rain is often accompanied by lightning and thunder. In tropical latitudes this type of rain is usually torrential. It is the most common type of rain to fall in equatorial regions and in regions having a Tropical Monsoon Climate.

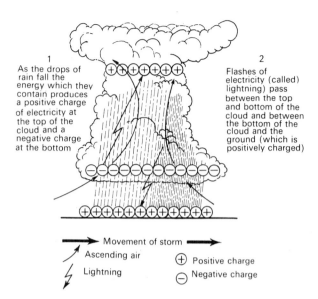

1 As the drops of rain fall the energy which they contain produces a positive charge of electricity at the top of the cloud and a negative charge at the bottom

2 Flashes of electricity (called) lightning) pass between the top and bottom of the cloud and between the bottom of the cloud and the ground (which is positively charged)

→ Movement of storm ➤

↗ Ascending air

⚡ Lightning

⊕ Positive charge

⊖ Negative charge

Thunderstorms can occur whenever land surfaces become greatly heated. In humid tropical regions like Indonesia, Malaysia, Central and West Africa, the Amazon Basin and Central America, thunderstorms are very common. They usually occur in the afternoon and are especially frequent in the season of heavy convectional rains.

The thunder of these storms is caused by the rapid expansion and contraction of the air. The electrical discharges (lightning) produce intense heat which causes the air to expand. Cooling soon takes place and the air contracts.

II Depression or Cyclonic Rain

Depression rain occurs when large masses of air of different temperatures meet. The warm air is forced up and over the cooler air. In tropical cyclones the rainfall is often very heavy but lasts for only a few hours. In temperate depressions it is much lighter but lasts for many hours or even days. Cyclonic rain is common throughout the Doldrums where the trade winds meet.

Warm air rises over cold air. It expands and cools. Condensation takes place and clouds and rain form.

This line represents the plane separating warm air from cold air

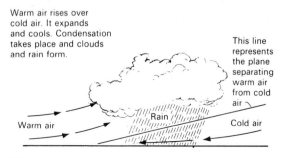

Warm air   Rain   Cold air

III Orographic or Relief Rain

Whereas convection rain only occurs in regions whose surfaces are greatly heated by the sun, and cyclonic rain only occurs where masses of air of different temperatures meet, orographic rain occurs in all latitudes. It is most common where on-shore winds rise up and over hilly or mountainous regions lying at right angles to the direction of the winds.

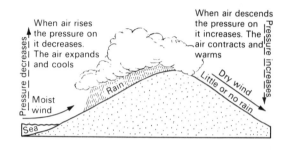

When air rises the pressure on it decreases. The air expands and cools

Pressure decreases

Moist wind

Sea

Rain

When air descends the pressure on it increases. The air contracts and warms

Pressure increases

Dry wind

Little or no rain

Fogs

Extensive fogs develop where warm and cold currents meet and where warm moist winds blow over cold surfaces. Fogs are very frequent off the mouth of the St. Lawrence (map below) and round the shores of Japan where the warm Kuro Siwo and the cold Oya Siwo (Okhotsk) meet. They often develop off the west coasts of hot deserts where warm sea breezes pass over the cold off-shore current. The sea breezes are only local winds but they give rise to belts of mist or fog just off the coasts. Examples of this type occur off the coasts of California, Peru, N.W. Africa and S. Africa (lower map).

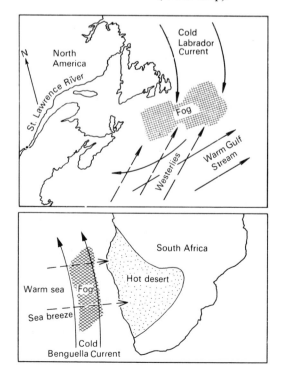

North America

St. Lawrence River

Cold Labrador Current

Fog

Westerlies

Warm Gulf Stream

South Africa

Hot desert

Warm sea   Fog

Sea breeze

Cold Benguella Current

Mean annual rainfall for the world

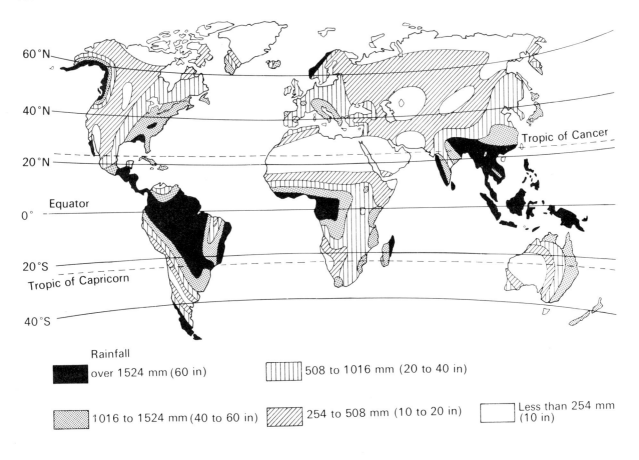

Rainfall

■ over 1524 mm (60 in)

▦ 508 to 1016 mm (20 to 40 in)

▨ 1016 to 1524 mm (40 to 60 in)

▨ 254 to 508 mm (10 to 20 in)

☐ Less than 254 mm (10 in)

The World Pattern of Rainfall

We have already studied how the revolution of the earth and the tilt of its 'axis' result in a movement of some pressure belts and a change in others, and how this in turn causes the belts of prevailing winds to move northwards and southwards, and other wind belts to change direction during the year. Let us now see how this affects the rainfall pattern of the world. First, we will examine the map at the top of this page which shows the mean annual rainfall. In this map the world has been divided into five types of rainfall region on a basis of the amount of rain received annually. Notice the following points.

1 The wettest regions (over 1500 mm about 58 in) occur in:
   (i) Equatorial Latitudes (Amazon and Zaire Basins, Indonesian Islands, Malaysia and New Guinea)
   (ii) Tropical Monsoon Regions (S. China, Peninsular S.E. Asia, Bangladesh (formerly E. Pakistan), N.E. India and W. India, Sri Lanka (Ceylon) and the Philippines)
   (iii) Regions receiving on-shore Westerly Winds (British Columbia, N.W. Europe, S. Chile, Tasmania, South Island of New Zealand).

2 The driest regions (less than 250 mm about 10 in) occur in:
   (i) Hearts of N. America and Asia
   (ii) Regions lying permanently under off-shore Trade Winds (Sahara and Arabian Deserts, Australian Desert, Kalahari and Atacama Deserts and the deserts of S.W. States of U.S.A.)
   (iii) Arctic Lowlands (N. America, Greenland and Asia).

Now this map only tells us how much rain a region receives each year. It does not tell us at what time of the year the rain comes. When we study agriculture, i.e. the types of crops grown on the earth's surface, it is necessary to know both the amount of rain falling in a region in one year and the time of year when it comes. The maps on page 142 and 143 show the distribution of rain for the world during the summer and winter seasons.

May 1st to October 31st

(See map on page 129)

1 The sun is overhead in the Northern Hemisphere and most of the rain falls in this hemisphere.

2 The belt of equatorial convection rains is chiefly located north of the equator.

3 Southern and eastern Asia and eastern N. and S. America receive heavy rain from on-shore winds.

4 Extensive areas in S.W. Asia, N. Africa, N. and Western Australia, S. Peru, N. Chile, S.W. States of N. America, and the Namib Desert receive little or no rain because they lie under off-shore trade winds.

5 The Arctic lowlands receive little rain because of low temperatures which prevent the air from absorbing much water vapour.

November 1st to April 30th
(see map on page 143)

1 The sun is overhead in the Southern Hemisphere and most of the rain falls in this hemisphere.

2 The belt of equatorial convection rain is chiefly located south of the equator.

3 N. and E. Australia, S.E. Africa and S.E. Brazil and E. Argentina receive rain from on-shore trade winds. (In N. Australia rain is brought by monsoon winds which are modified N.E. Trade Winds.)

4 Extensive areas of S.W. Asia, N. Africa, Central and W. Australia, S. Peru, N. Chile and S.W. Africa receive little or no rain because they lie under off-shore trade winds.

5 The Arctic Lowlands as before have little or no rain.

The Migration of the Overhead Sun and the Rainfall Pattern

A *Some regions are NOT influenced*

1 Regions permanently in a belt of prevailing winds:
   (i) N.W. Europe, W. Canada, S. Chile, Tasmania, South Island (N.Z.).
   They lie in the belt of on-shore Westerlies (rain all the year).
   (ii) S. California, N. Chile, Sahara, S.W. Asia, W. Australia and the Namib Desert.
   They lie in the belt of off-shore Trades (little or no rain throughout the year).

2 Regions permanently in the doldrums belt, e.g. Zaire and Amazon Basins. They receive convection rain throughout the year.

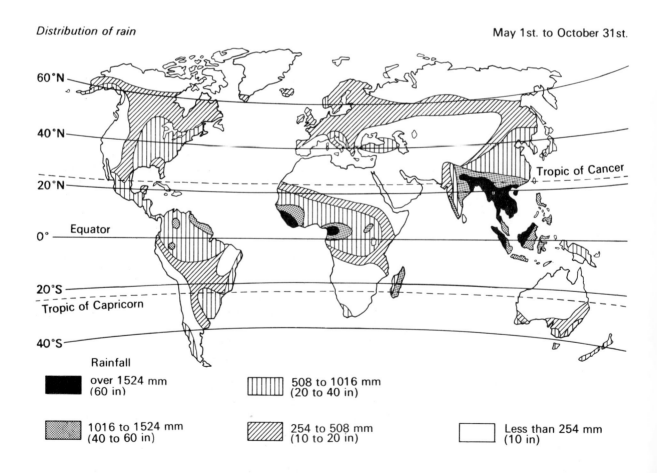

Distribution of rain   May 1st. to October 31st.

Rainfall

█ over 1524 mm (60 in)

▦ 1016 to 1524 mm (40 to 60 in)

▥ 508 to 1016 mm (20 to 40 in)

▧ 254 to 508 mm (10 to 20 in)

☐ Less than 254 mm (10 in)

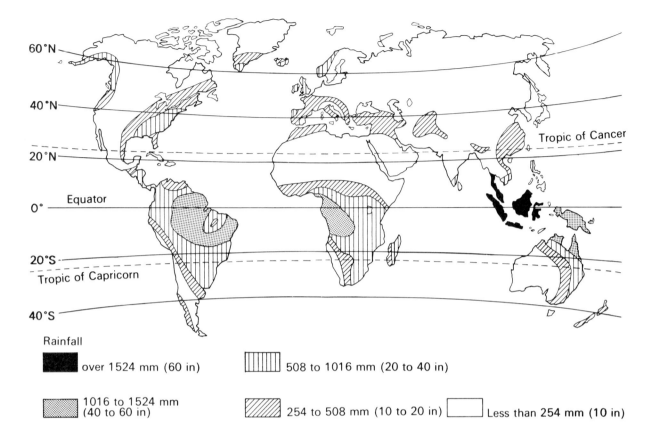

Rainfall

- ■ over 1524 mm (60 in)
- ▦ 508 to 1016 mm (20 to 40 in)
- ▧ 1016 to 1524 mm (40 to 60 in)
- ▨ 254 to 508 mm (10 to 20 in)
- □ Less than 254 mm (10 in)

B *Some regions ARE influenced*

1 Regions lying between two belts of prevailing winds:
Central California, Central Chile, Mediterranean Lowlands, S.W. Australia.
They lie between the Westerly and Trade Wind belts. The Westerlies bring rain in the cool season; in the warm season the Trades blow and there is no rain.

2 The interiors of Asia and N. America. In these regions atmospheric pressure is low in summer. In-blowing winds give rain and intense heat causes convection rain. In winter atmospheric pressure is high and winds are out-blowing. There is little or no rain.

3 Regions bordering the permanently wet equatorial regions lie under the doldrums once a year and this results in heavy convection rain. They also lie under the Trades once a year. Where these are off-shore there is little or no rain. These regions lie in S. America, Africa and Australia and they have a *Sudan climate* (see page 150).

4 Monsoon Regions:
Japan, China, Peninsular S.E. Asia, India, Bangladesh, Sri Lanka and N. Australia.
For one season they lie under on-shore winds which bring rain. For the other season they lie under off-shore winds which bring little or no rain.

In the map on page 144 the world has been divided into four types of rainfall region. The four types are:
 (i) Regions receiving rain throughout the year
 (ii) Regions receiving rain chiefly in the hot season (summer)
 (iii) Regions receiving rain chiefly in the cool season (winter)
 (iv) Regions receiving little or no rain.

The following statements can be made about this map:

1 *Regions having rain all the year are located in*:
 (i) Equatorial and some tropical latitudes
 (ii) Coastal regions where winds are on-shore for most of the year.

2 *Regions having most rain in the summer are located in*:
 (i) Most of Asia excluding Malaysia and In-

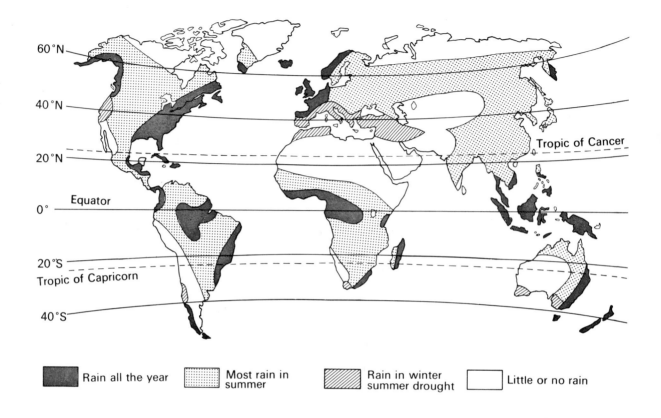

Rain all the year Most rain in summer Rain in winter summer drought Little or no rain

donesia and New Guinea
(ii) E. Europe
(iii) N. Australia
(iv) Regions bordering the equatorial latitude of Africa and S. America
(v) Central N. America.

3 *Regions having most rain in the winter are located in*:
(i) Central California and Central Chile
(ii) Mediterranean Lowlands
(iii) S.W. Africa and S.W. Australia.

4 *Regions having little or no rain are located in*:
(i) S. California and S.W. States of N. America
(ii) The Atacama and Kalahari Deserts
(iii) The Sahara, Arabian and Asian Deserts
(iv) The Australian Desert
(v) The Polar Deserts.

Seasonal Rainfall and Type of Rain

On page 145 is a diagrammatic representation of Eurasia and N. Africa and it shows the seasonal distribution of winds and rain together with the type of rain The winds on the left-hand side operate

over the western part of the region and those on the right-hand side operate over the eastern part of the region. Examine this figure and pay particular attention to the following:
(i) The north-south shift of some wind belts, and the change in wind direction of other belts from season to season.
(ii) The location of on-shore winds (which usually bring rain) and of off-shore winds (which usually bring no rain).
(iii) The distribution of rain in relation to the lines of latitude which are given.

Recording of Rainfall on a Map

Lines are drawn on the map through all places having the same rainfall. Such lines are called *isohyets*. They are drawn at a uniform interval (in the lower left diagram on page 145 this is 25). A scale of colours or line shading (see lower right diagram on page 145) is then worked out and applied to the map. It is usual to start with light colours or open lines for low rainfall and to use darker colours or closer lines for heavy rainfall.

Rainfall maps may show seasonal or annual values.

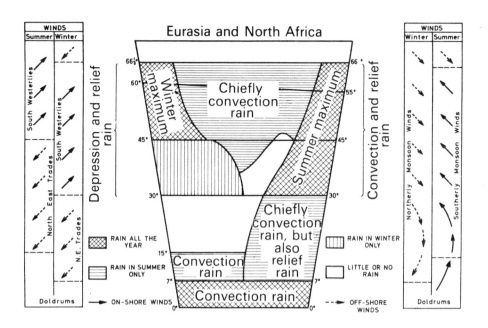

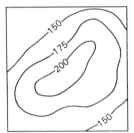

Rainfall in millimetres

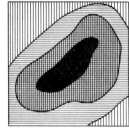

EXERCISES

1 Carefully explain the meanings of the following:
   (i) sea breezes are day winds, and land breezes are night winds
   (ii) water surfaces gain and lose heat more slowly than land surfaces
   (iii) temperature decreases as altitude increases
   (iv) the height of the mid-day sun in the tropics is always higher than it is in temperate latitudes. Illustrate your answer with well-labelled diagrams.

2 By using well-labelled diagrams, describe *three* ways by which rain may be caused, and for each, name a specific region where this type of rain commonly occurs.

3 Explain each of the following statements:
   (i) most of the hot deserts are located on the western sides of continents between 20°S. and 30°S. and 20°N. and 30°N.
   (ii) heavy fog frequently occurs over the waters around Newfoundland

(iii) the surface waters of the north-east Atlantic are warmer than those of the north-west Atlantic

4 Carefully explain the meaning of each of the following:
   (i) mean monthly temperature
   (ii) annual temperature range
   (iii) mean annual temperature
   (iv) diurnal temperature range

5 Briefly discuss the different factors which affect the world temperature distribution pattern, and explain the part played by temperature in climate.

6 Explain the following statements:
   (i) water vapour and dust influence the amount of insolation a region receives
   (ii) descending winds are warm winds
   (iii) temperate cyclones give changeable weather
   (iv) extremes of temperature often occur in the hearts of continents

7 Name *four* local winds and for *each*:
   (i) describe why it develops
   (ii) explain why it is warm or cold, dry or damp
   (iii) name a region where it operates

8 Name the instruments used for measuring wind direction and velocity and state how these two characteristics can be shown in diagrammatic form.

9 Carefully explain, with the aid of labelled diagrams, the following statements:
   (i) equatorial regions receive rain throughout the year
   (ii) the highest daily temperatures are recorded outside equatorial latitudes
   (iii) the hearts of the northern continents receive little rain.

Objective Exercises

1 The approximate amount of the Sun's energy (called insolation) which reaches the Earth's surface is
   A 90%
   B 75%
   C 45%
   D 30%
   E 20%

   A B C D E
   ☐ ☐ ☐ ☐ ☐

2 Four towns, all at about the same altitude and on the same line of latitude, have mean January temperatures as given below. Which town is farthest from the sea?
   A   7°C
   B −18°C
   C −10°C
   D   3°C
   E   10°C

   A B C D E
   ☐ ☐ ☐ ☐ ☐

3 The difference between the maximum and minimum temperatures recorded for a place during a period of one day is called
   A daily mean temperature
   B diurnal temperature range
   C daily average temperature
   D mean monthly temperature
   E mean annual temperature

   A B C D E
   ☐ ☐ ☐ ☐ ☐

4 Perpendicular rays are usually more heating than oblique rays. It can therefore be said that
   A a south-facing slope is warmer than a north-facing slope in the summer in the Northern Hemisphere
   B a north-facing slope is warmer than a south-facing slope in the summer in the Northern Hemisphere
   C the seasons in the tropics are short
   D winter in latitude 35°S is colder than winter in latitude 35°N
   E summer temperatures in central Asia are lower than they are in central Africa

   A B C D E
   ☐ ☐ ☐ ☐ ☐

5 Which of the following winds are predominantly seasonal winds?
   A depression(cyclonic) winds
   B prevailing winds
   C monsoon winds
   D local winds
   E descending winds

   A B C D E
   ☐ ☐ ☐ ☐ ☐

6 All of the following are cyclones except
   A hurricane
   B Zonda
   C Willy Willy
   D typhoon
   E tornado

   A B C D E
   ☐ ☐ ☐ ☐ ☐

7 Which one of the following is not a necessary condition for the formation of heavy rainfall?
   A air must be warm near the surface
   B moist air must rise to great heights
   C air must become saturated
   D air pressure must be high
   E humidity must be high

   A B C D E
   ☐ ☐ ☐ ☐ ☐

8 Which one of the following statements relates to both land and sea breezes?
   A Air blows from the sea to the land during the day.
   B Air blows from the land to the sea during the night.
   C Air generally moves from a cool region to a warmer region.
   D The land cools more quickly than the sea during the night.
   E The sea warms up more slowly than the land during the day.

   A B C D E
   ☐ ☐ ☐ ☐ ☐

9 All of the following statements are true except
   A Relative humidity of a mass of air falls if the temperature of the air rises.
   B Air is saturated when its relative humidity is 100%.
   C When air subsides its relative humidity decreases.
   D The relative humidity of a mass of air remains constant when the air crosses over a cold land surface from a warm water surface.
   E Condensation takes place in a mass of air if there is further cooling after the relative humidity of the air reaches 100%.

   A B C D E
   ☐ ☐ ☐ ☐ ☐

10 A line on a weather map joining all places of equal pressure is called an
   A isotherm
   B isohyet
   C isohel
   D isobar

   A B C D
   ☐ ☐ ☐ ☐

11 Which one of the following statements most accurately describes the relationship between air pressure and air temperature?
  A When air is heated it becomes lighter and it rises.
  B Air which is cooled contracts and the outward pressure of its molecules is spread over a smaller area.
  C When air rises its pressure decreases.
  D The pressure increases in a body of air which is descending.

  A B C D
  ☐ ☐ ☐ ☐

12 All of these statements are true about anticyclones in general but only one is true of anticyclones in the Southern Hemisphere. Which is that statement?
  A The air moves in a circular manner.
  B Pressure increases from the outside to the centre.
  C Anticyclones often form over the continents.
  D The air moves in an anti-clockwise direction.

  A B C D
  ☐ ☐ ☐ ☐

13 The land on the leeward side of mountain ranges which are at right angles to on-shore winds is often dry. This is because
  A the winds are descending on the leeward side
  B pressure is high to the leeward side
  C the air on the leeward side is cool and is therefore relatively dry
  D the leeward side lies under dry land winds

  A B C D
  ☐ ☐ ☐ ☐

14 Equatorial lowlands usually experience
  A a large diurnal temperature range
  B heavy thunder rain in the afternoon
  C strong winds
  D cold nights

  A B C D
  ☐ ☐ ☐ ☐

15 The seasonal rainfall pattern of India is caused by
  A the large annual range of temperature
  B the tropical location of India
  C the monsoon winds operating over southern Asia
  D the Himalayas blocking winds from interior Asia

  A B C D
  ☐ ☐ ☐ ☐

16 A line on a map which joins places having the same rainfall is called an
  A isohyet
  B isobar
  C isotherm
  D isohel

  A B C D
  ☐ ☐ ☐ ☐

17 Altitude is one of several factors which influence the temperature of a place. Which one of the following statements best explains the influence of altitude on temperature?
  A A place which has a high altitude is nearer the Sun and it therefore has a higher temperature than it would have at sea level.
  B Because the Sun's rays meet the top levels of the atmosphere first, high altitudes tend to be warmer than low altitudes.
  C The atmosphere is heated primarily from below and therefore temperatures at high altitudes are lower than temperatures at low altitudes.
  D Air is rarefied at high altitudes and it contains very little vapour or dust. Heat from the Earth's surface therefore rapidly escapes and temperatures at high altitudes are lower than they are at low altitudes.

  A B C D
  ☐ ☐ ☐ ☐

18 Which one of the following factors can have the greatest influence on the temperature of a place in Equatorial latitudes?
  A aspect
  B distance from the sea
  C altitudes
  D ocean currents

  A B C D
  ☐ ☐ ☐ ☐

19 Which one of the following statements about winds and atmospheric pressure is **not** correct?
  A Winds blow in a clockwise direction around a centre of low pressure in the Northern Hemisphere.
  B Winds blow in a clockwise direction around a centre of high pressure in the Northern Hemisphere.
  C Winds blow in an anti-clockwise direction around a centre of high pressure in the Southern Hemisphere.
  D Winds blow in a clockwise direction around a centre of low pressure in the Southern Hemisphere.

  A B C D
  ☐ ☐ ☐ ☐

20 The windward slopes of coastal mountains which are at right angles to winds blowing from the sea are wetter than the leeward slopes. This is because
  A they are nearer the sea
  B the winds have to rise to cross them
  C descending winds are warm
  D the sea is warmer than the land

  A B C D
  ☐ ☐ ☐ ☐

21 If the temperature at sea level is 7°C, the temperature of the air at a height of 2000 metres will be about
  A 15°C
  B 16°C
  C −6°C
  D 10°C

  A B C D
  ☐ ☐ ☐ ☐

14   Types of Climate

All of us know that some regions are hot and others are cold; some are wet and others are dry; some have rain all the year and others have rain for part of the year only.

The world can be divided into climatic regions, each of which has a distinct temperature and rainfall pattern. This can be done by dividing the world into five temperature zones:

Hot Zone; Warm Zone; Cool Zone; Cold Zone; Very Cold Zone.

Each of these zones is very large, and, with the exception of the very cold zone, the rainfall* distribution is not even, i.e. one part of a zone may be wet whilst another part of the same zone may be dry. All except the very cold zone are now subdivided into rainfall regions. The resulting regions all have a distinct climate and these are shown in the map below. The only climatic type shown on this map which is not based on the temperature and rainfall division is the mountain climate. This map shows 18 climatic types in all and you must become familiar with all of them.

The main features of each climatic type are summarised in three or four statements, and a climatic diagram is drawn for a particular town for each type. At least two towns have been selected for each of the tropical climates. The climatic diagrams show:

1 Monthly Rainfall
2 Total Annual Rainfall
3 Annual Temperature Range

TROPICAL CLIMATES

EQUATORIAL CLIMATE (Padang and Singapore)

Location

1 The best examples occur in the lowlands between

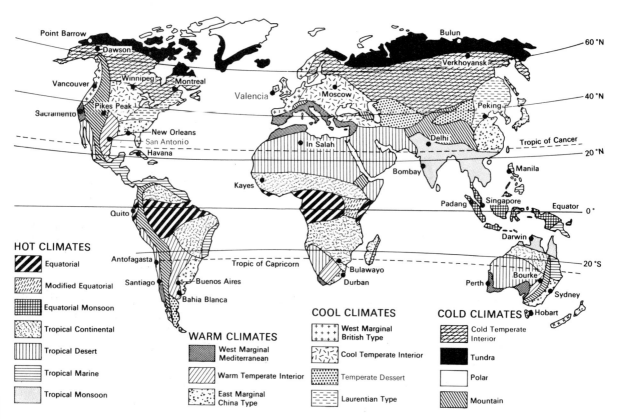

HOT CLIMATES
- Equatorial
- Modified Equatorial
- Equatorial Monsoon
- Tropical Continental
- Tropical Desert
- Tropical Marine
- Tropical Monsoon

WARM CLIMATES
- West Marginal Mediterranean
- Warm Temperate Interior
- East Marginal China Type

COOL CLIMATES
- West Marginal British Type
- Cool Temperate Interior
- Temperate Dessert
- Laurentian Type

COLD CLIMATES
- Cold Temperate Interior
- Tundra
- Polar
- Mountain

Precipitation covers both rain and snow, and this word should be used for the very cold climates and other climates where snowfall is significant.

5°N. and 5°S., e.g. the Amazon and Zaire Basins.

2 The highlands which occur between these latitudes, e.g. East African highlands, have a much modified equatorial climate. Altitude in these regions 'reduces' temperatures to about 15°C (59°F).

3 A part of the Guinea coast of West Africa receives low annual rainfall, e.g. Accra, 700 mm (28 in). This region really has an equatorial climate which is modified by monsoon winds.

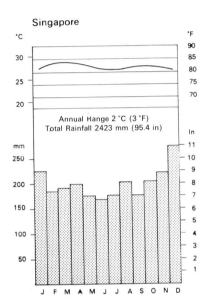

Singapore

Annual Range 2 °C (3 °F)
Total Rainfall 2423 mm (95.4 in)

Climatic characteristics

1 This latitudinal belt lies under the Doldrums Low Pressure all the year and, therefore, there are no seasons.

2 The mid-day sun is always near the vertical and it is overhead twice a year, at the equinoxes.

3 Average daily temperatures are 26°C (about 79°F) throughout the year. These are well below the average daily temperatures of other types of climate occuring outside equatorial latitudes. Extensive cloud cover and heavy rainfall prevent temperatures from rising much over 26°C (about 79°F).

4 The diurnal temperature range is between 6°C (11°F) and 8°C (15°F) which is greater than the annual temperature range of about 3°C (5°F).

5 Rainfall is heavy and is usually convectional. Rains often come in the afternoons and are generally accompanied by lightning and thunder. Annual rainfall is about 2000 mm (80 in) though some regions get higher falls.

Note: Malaysia, Singapore and Indonesia are generally said to have an equatorial climate. Although these countries show some of the characteristics of this climate, e.g. the temperature and

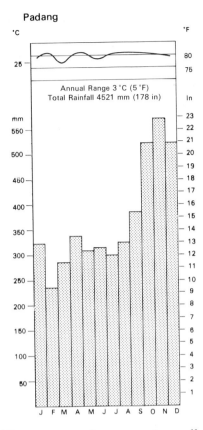

Padang

Annual Range 3 °C (5 °F)
Total Rainfall 4521 mm (178 in)

humidity patterns, others are not as well developed. This is because they lie under the monsoon winds which sweep across South-East Asia from the N.E. from November to February and from the S.W. from July to September. These winds moderate humidity and give a seasonal pattern to the rainfall regime of many regions.

6 Humidity is always high.

Agricultural development

1 In some of the more heavily forested parts of equatorial regions, populations are sparse and hunting and food collecting form the basic economies of the more primitive people, while shifting cultivation is practised by the more advanced forest dwellers.

2 Plantation agriculture for the cultivation of rubber (Malaysia and Indonesia), cacao (West Africa, especially Ghana) and oil palm (Nigeria and Malaysia) and other crops like sugar, bananas, and pineapples, etc. has become established in many coastal regions.

3 Equatorial forests contain valuable trees such as mahogany, ebony and greenheart, and the cutting down of these in some regions has given rise to important timber industries.

Extensive development is often made difficult by many factors, the most important of which are:

(i) diseases and insect pests which attack man, his animals and his crops, (ii) the difficulty of establishing communications in the very densely forested regions, (iii) the generally poor, thin soils which soon lose what little goodness they contain when the forest cover is removed.

TROPICAL CONTINENTAL (SUDAN) CLIMATE
(Kayes and Bulawayo)
Location
1 It occurs between 5°N. and 15°N., and 5°S. and 15°S.
2 It is best developed in Africa and east central South America.

Climatic characteristics
1 The latitudinal belt which has this climate comes under the Trade Winds for a part of the year (winter) and lies under the doldrums for the rest of the year (summer).
2 Summers are hot with temperatures around 32°C (about 90°F). Winters are cooler, 21°C (about 70°F). The annual temperature range is therefore about 11°C (20°F).
3 Heavy rains, mainly of convectional type, fall in in the summer: winters are usually dry.
4 The Trade Winds in north Africa blow from the Sahara Desert and are dry, hot winds. These are particularly noticeable in West Africa where they are called the *harmattan*. Besides being dry and hot, the harmattan are also dusty. In South Africa and in South America, south of the equator, the Trade Winds blow from the sea and they bring rain to coastal regions.
5 The annual rainfall is often around 762 mm (30 in) but it may be more, e.g. for regions near to the equatorial latitudes, or less e.g. regions near the hot deserts.

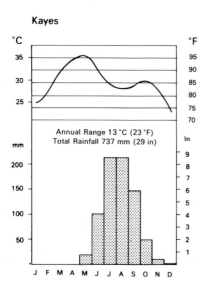

Kayes

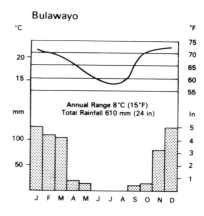

Bulawayo

6 The highest temperatures occur just before the rainy season begins, i.e. April in the Northern Hemisphere and October in the Southern Hemisphere
7 Humidity is high in the summer.
8 The climate can be said to have hot, wet summers and cooler, drier winters, but note the exceptions referred to in (4).

Agricultural development
The natural vegetation consists of tall grass, often 3 m (10 ft) or more, with scattered clumps of trees. In Africa this type of vegetation is called *savana*, and it is the home of a great variety of animals such as elephant, giraffe, zebra, deer, leopard, tiger jaguar, etc.

1 Agricultural development is not far advanced in the African savanas. The Masai, who are nomadic herdsmen, maintain large herds of zebu cattle, goats and sheep which graze on the grasslands of Kenya, Tanzania and Uganda. The cattle are reared for their milk and blood both of which form important items in the diet of the Masai. The blood is taken from the neck of the animal which does not appear to suffer any ill effects. The cattle are not killed for meat.
2 Crops are grown in the African savanas, especially in northern Nigeria where the Hausa have practised crop cultivation for a long time. They grow food crops such as Guinea corn, millet, maize, bananas, groundnuts and beans. They also grow non-food crops such as cotton and tobacco which are sold for cash. The Hausa practise crop rotation which enables soil fertility to be maintained. They also keep herds of goats and cattle which supply them with milk and meat.
3 Similar crops are cultivated by the Kikuyu in East Africa and some of their produce is sold to the Masai.
4 Commercial plantation farming has become established in some parts of the African savana, especially in East Africa, Uganda, Malawi, Kenya

and Tanzania. The main crops of these plantations are sugar cane, tobacco, sisal and cotton.

5 The two savana regions of South America, known as the *llanos* and the *campos*, are both used for rearing cattle. The latter is the more extensive and the more important for animal farming. Its potentials are considerable.

6 The savana of Australia is very sparsely populated and as yet there has been no significant agricultural development.

The main factors which at present prevent any further expansion to agricultural development are (i) the unreliable nature of the rainfall (droughts are common), (ii) diseases and insect pests which attack animals and crops, (iii) loss of soil fertility through the removal of such minerals as phosphates, potash and nitrates by heavy rain in the west season, (iv) poor communications, and (v) a sparse population in relation to the extent of the savana lands. It is possible that all of these conditions will be put right or counteracted eventually, and if and when this happens, the savana lands will become major producers of animal products, and food and non-food crops.

TROPICAL MARINE CLIMATE (Havana, Durban and Manila)
Location
1 It occurs in east coastal areas of regions having a tropical continental climate, and in the coastal regions of Central America.
2 It is developed best in the lowlands of Central America, the West Indies, coastal lowlands of Brazil and East Africa, including east Malagasy, north-east Australia and the Philippines.

Climatic characteristics
1 On-shore trade winds blow throughout the year and they bring rain almost every day with rather heavier falls in the hot season.

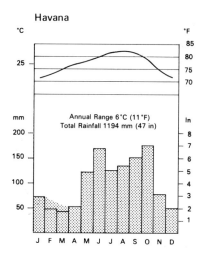

2 Annual rainfall varies from 1000 mm to 2000 mm (40 in to 80 in) depending upon location. Rainfall is both convectional and orographic.

3 Temperatures are similar to those of tropical continental climates. The annual temperature range is about 8°C (15°F) with hot season temperatures of 29°C (about 85°F) and cool season temperatures of 21°C (about 70°F).

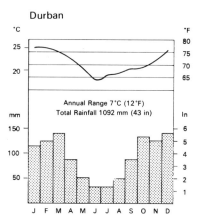

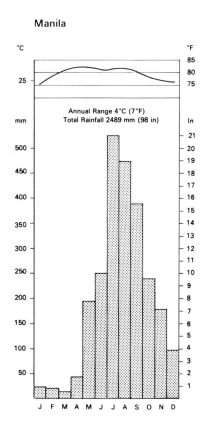

151

Agricultural development

1 The cultivation of crops, especially food crops is important in almost all tropical marine regions. Sugar-cane is grown annually on large plantations in most parts of the West Indies, East Africa (especially Natal) and Queensland (north-east Australia). Other crops include coffee in eastern Brazil, and Manila hemp in the Philippines.

TROPICAL MONSOON CLIMATE (Delhi, Bombay and Darwin)

Location

In some tropical and temperate latitudes seasonal land and sea winds operate on a huge scale affecting both continents and oceans. These seasonal winds, called *monsoon winds,* are best developed over an area extending from south-eastern and eastern Asia to northern Australia. Eastern Asia has a temperate monsoon climate (refer to the maps on page 131).

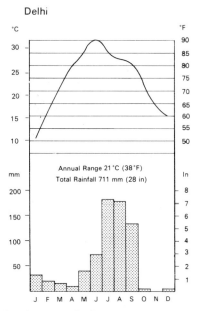

Delhi

Annual Range 21 °C (38 °F)
Total Rainfall 711 mm (28 in)

Climatic characteristics

1 Seasonal reversal of winds is the chief feature of this climate. For one season the winds blow from the sea to the land bringing heavy rainfall to coastal regions, and for another season the winds blow from the land to the sea and these give little or no rain.

2 Annual rainfall varies greatly, the amount falling depending mainly on relief and the angle at which the on-shore winds meet this. The south-facing slopes of the Khasi Hills in Assam receive as much as 12 500 mm (500 in) annually. In contrast the region around Delhi receives only 620 mm (25 in) annually.

3 Temperatures range from 32°C (about 90°F) in the hot season to about 15°C (59°F) in the cool

season, thus giving an annual range of about 17°C (31°F). But these values vary very much (compare Bombay and Delhi temperature graphs).

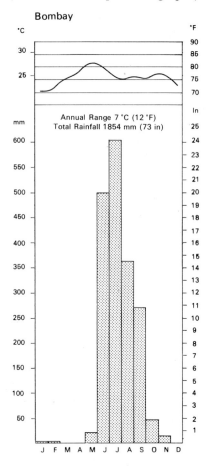

Bombay

Annual Range 7 °C (12 °F)
Total Rainfall 1854 mm (73 in)

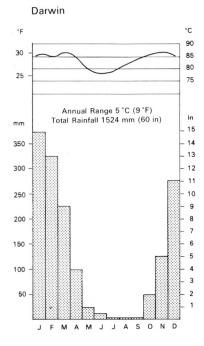

Darwin

Annual Range 5 °C (9 °F)
Total Rainfall 1524 mm (60 in)

4 A typical tropical monsoon climate consists of three seasons. For the Indian sub-continent and Burma these are:
   (i) Cool, dry season (November to February) when the off-shore north-east monsoon winds blow.
   (ii) Hot, dry season (March to May) when temperatures are high because of the near overhead mid-day sun, and when winds are almost absent.
   (iii) Hot, wet season (June to October) when the on-shore south-west monsoon winds blow and when rainfall is very heavy in regions receiving the full force and fury of on-shore winds. During this season cloudy skies cause temperatures to fall a little, but humidity rises to its maximum for the year.

Agricultural development
1 The intensive cultivation of food crops takes place in all parts of tropical monsoon regions where soils are adequately fertile and rainfall or the availability of water in the form of lakes and rivers, is sufficient to meet crop requirements.
2 There is also extensive cultivation of non-food crops usually on plantations or estates which in the main have been established on cleared forest land.
3 Padi is the most commonly cultivated food crop and it forms the staple diet of most of the people in the wetter parts of India, Pakistan, Burma, southern China, and the countries of northern South-East Asia. Wet padi is grown on well-watered lowlands and hill slopes (which have to be terraced). Dry padi, which does not require as much water, is grown on drier hill slopes.
4 Wheat, millet, maize, sorghum, etc. are grown in the drier areas where padi cannot be grown. These are particularly important food crops in northern India and Pakistan.
5 Other important lowland crops are sugar cane, especially in the wetter regions, cotton, a most important crop in India and Pakistan, and jute, mainly in the Ganges Delta.
6 In the highland areas of many tropical monsoon regions tea forms an important plantation crop. This crop is extensively grown in Sri Lanka and the Himalayan foothills of India and Bangladesh.

The cultivation of food crops, especially padi is entirely dependent upon the on-shore monsoon bringing the right amount of rain at the right time. If the monsoon arrives late it can result in widespread famine. Alternatively, monsoon rains may be unusually severe, resulting in widespread flooding and crop destruction, thus again causing famine.

The main obstacle to better farming in many tropical monsoon regions is the farmers' ignorance of modern farming methods involving maintenance and improvement of soil fertility by using fertilisers, manure and crop rotation. Before this obstacle can be overcome, the practice of land division on the death of the owner will probably have to be stopped, and money will probably have to be made available to farmers for the purchase of good seed, fertilisers and labour-saving farming equipment.

TROPICAL DESERT CLIMATE (Antofagasta and In Salah)

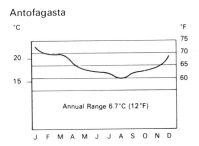

Location
1 It occurs on the western sides of land masses in the belt of permanent trade winds. The only exception is the desert belt in north Africa which extends from the west coast right across the continent and into south-west Asia. This is because the trade winds affecting the eastern part of north Africa are blowing in from the land mass of south-west Asia and they are therefore dry winds.
2 The most important regions having this type of climate are the Sahara Desert, the Arabian, Iranian and Thar Deserts, the Australian Desert, the Kalahari and Namib Deserts, the Atacama Desert and the Californian and Mexico Deserts.

Climatic characteristics
1 Rain rarely falls and the average annual fall is usually below 12 cm (5 in). Sometimes there are sudden torrential downpours which give rise to temporary flooding.
2 The hot deserts occur in the tropical high pressure belts where air is subsiding. Such air absorbs rather than yields moisture. Also, most of the winds blowing into the hot deserts originate in cooler regions. When crossing the desert these winds get hotter and again this prevents condensation. Note also that on-shore winds which sometimes affect the west coasts of desert belts cross cold currents which occur along these coasts. The winds are cooled and condensation takes place giving fog or light showers. The winds are therefore dried and on reaching the warm land surface they are dried still further.

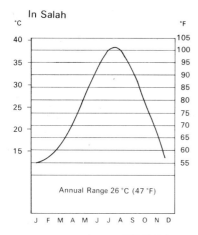

In Salah

Annual Range 26 °C (47 °F)

3 Temperatures vary from 29°C (about 85°F) in the hot season to 10°C (50°F) in the cool season.
4 Because there are no clouds, day temperatures often go over 38°C (about 100°F) and temperatures of over 49°C (120°F) are not uncommon. At night, again because there is no cloud cover, radiation is rapid and temperatures can fall to 15°C (59°F) in the hot season to 5°C (about 40°F) in the cool season. Diurnal temperature ranges are therefore very high.

Agricultural development
1 Regions having this type of climate can be cultivated and lived in if there is a constant supply of water.
2 The lowlands of the valleys of the Nile, the Tigris-Euphrates and the lower Indus are intensively cultivated for both food and non-food crops. By developing irrigation canals, all of these regions have extended the area of cultivation beyond the limits of the river valleys.
3 The cultivation of crops in desert areas where there are no rivers can only take place if water is available in some other form. Water does occur in oases and it is at these sites that settled agriculture takes place. Some oases are small but some are very extensive. Dates, wheat, vegetables and fruit are usually grown. Oases occur mainly in the Sahara and Arabian Deserts.
4 Nomadic herding takes place in the Sahara and Arabian Deserts where Bedouin Arabs own large flocks of sheep and goats. These animals graze on the scant pastures which occur in some parts of the desert, especially in the higher parts where there is heavy condensation,.in the form of dew at night. The animals provide the Bedouin with most of his basic requirements: the rest he obtains by barter trade with the oases farmers.

The main factor preventing further development in the deserts is the very limited quantities and distribution of available fresh water. Some deserts, especially the Sahara Desert, are known to have large supplies of water below the surface. One day it may be possible to make this available on a huge scale which could then see an extension of farming activities. It may also be possible to tow ice-bergs from the Antarctic to the hot desert areas.

WARM TEMPERATE CLIMATES
WARM TEMPERATE WESTERN MARGIN CLIMATE – MEDITERRANEAN CLIMATE
(Sacramento, Santiago and Perth)

Location
1 This occurs between 30°N. and 45°N. and 30°S. and 40°S., on the western sides of continents.
2 The climate is best developed around the shores of the Mediterranean Sea, in south-west Africa, central Chile, central California, and south-west and southern Australia (Adelaide to Melbourne).

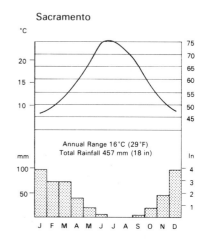

Sacramento

Annual Range 16°C (29°F)
Total Rainfall 457 mm (18 in)

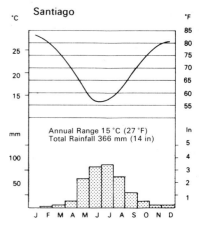

Santiago

Annual Range 15 °C (27 °F)
Total Rainfall 366 mm (14 in)

Climatic characteristic
1 Temperatures range from 21°C (about 70°F) in the summer to 10°C (50°F) or below in the winter.
2 Off-shore trade winds blow in the summer. These are dry and give no rain. The sky is cloudless and humidity is low.
3 On-shore westerly winds blow in the winter

bringing cyclonic rain. This usually amounts to about 500 mm (20 in) though considerably more may fall on steep slopes lying at right angles to the wind. The rain often comes in heavy showers which sometimes cause floods.

4 The annual rainfall ranges from 500 mm (20 in) to 760 mm (30 in).

5 Mediterranean climates experience both hot and cold local winds. Some examples are the *sirocco*, a hot dusty, dry wind which blows in the summer across the Mediterranean Sea from the Sahara Desert; the *mistral,* an intensely strong, cold wind, which blows in the winter down the Rhône Valley from the north and often reaching the Mediterranean coast; the *bora*, another winter

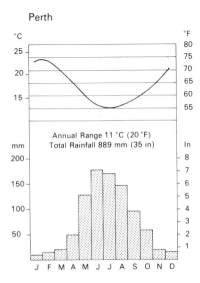

Perth

Annual Range 11 °C (20 °F)
Total Rainfall 889 mm (35 in)

wind which develops because of pressure differences between central Europe and the Mediterranean Sea and blowing south across Yugoslavia to the Adriatic.

6 The climate can be summarised as having bright, sunny, hot and dry summers and mild rainy winters.

Agricultural development

1 The climate permits a large range of crops to be grown. These include fruits and cereals.

2 Citrus fruits (oranges, lemons, grapefruit and limes) are extensively cultivated in many regions, very often by means of irrigation. The semi-desert regions of central California, central Chile and Israel are now very important producers of citrus fruits made possible by the use of irrigation. Other important fruits are peaches, apricots, plums, cherries and pears.

3 The olive tree is especially important to Mediterranean regions. Its fruit is rich in oil which is used in cooking in much the same way as coconut oil. The fruit is also eaten fresh and forms an

important item in the diet of Mediterranean people.

4 Nuts such as chestnuts, walnuts and almonds are grown.

5 The cultivation of grapes, which is called *viticulture*, takes place in most regions having a Mediterranean climate. A large part of the fruit is used for making wine, though some is dried (sultanas, currants and raisins) and some is eaten fresh.

6 The two most important cereals grown are wheat (the more important) and barley. Both crops are grown in the winter though in some regions, e.g. the Po Valley of Italy and the Ebro Valley of Spain they are grown in summer if irrigation water is available.

7 Other crops which are cultivated in some regions are figs, tobacco and cotton.

8 Agriculture has given rise to specialised industries such as wine-making, flour-milling, fruit-canning and food processing.

WARM TEMPERATE INTERIOR CLIMATE (Bourke and San Antonio)

Location

1 It occurs in the interior of continents, excluding Asia, between latitudes 20°N. and 35°N. and 20°S. and 35°S. The interior of Asia in these latitudes has very low winter temperatures and very low annual rainfall. Its climate is therefore better described as Temperate Desert.

2 The climate is best developed in the southern continents and is especially well developed in the Murray-Darling Lowlands (Australia) and in the High Veldt (Africa). It also occurs in the U.S.A. in western Oklahoma, Texas and northern Mexico.

Climatic characteristics

1 Temperatures range from 26°C (about 79°F) in the summer to 10°C (50°F) in the winter.

2 The annual rainfall varies from 380 mm (15 in) to 700 mm (30 in) depending on location. A good deal of the rain is brought by South-East Trade

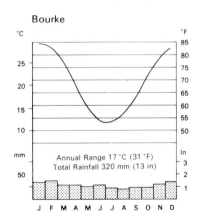

Bourke

Annual Range 17 °C (31 °F)
Total Rainfall 320 mm (13 in)

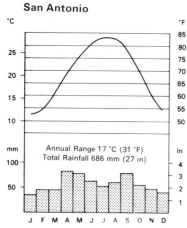

San Antonio

Annual Range 17 °C (31 °F)
Total Rainfall 686 mm (27 in)

Winds to the southern continental regions in the summer months. This decreases from east to west which means that the western parts of these climatic regions are fairly dry, sometimes semi-arid. There is also convectional rain which is caused by the low pressure systems over these areas in summer. Most of the rain occurs in summer.

3 Evaporation is high in summer and this often makes irrigation essential for crop cultivation.

Agricultural development

1 Grass is the natural vegetation of regions having this climate. The grasslands have specific names, e.g. the *Downs* (Australia), the *Veldt* (South Africa) and the *Pampas* (Argentina).

2 Vast herds of beef cattle and flocks of sheep are reared on these grasslands. Special fodder crops are grown for the animals.

WARM TEMPERATE EASTERN MARGIN CLIMATE – CHINA TYPE (New Orleans, Buenos Aires, Sydney)

Location

1 It is located on the eastern sides of continents between latitudes 23°N. and 35°N. and 23°S. and 35°S.

2 It is best developed in central China, south-eastern U.S.A., southern Brazil and the eastern part of the Pampas (Argentina), south-eastern Africa and south-eastern Australia.

Climatic characteristics

1 As in the Mediterranean climate, the trades and westerlies are the dominant seasonal winds. But notice the contrasts: in this climate the trades are on-shore winds and they bring rain whilst the westerlies are off-shore winds. These bring lighter rain.

2 Summers are hot 26°C (about 79°F) and winters are mild 13°C (about 55°F). Temperatures in the winter can be dramatically lowered suddenly

when local winds caused by depressions develop, e.g. *pampero* (Argentina) and *southerly burster* (Australia), and blow strongly.

3 Monsoonal winds tend to develop in both south-eastern U.S.A. and in China. In China the development is marked and there is a definite seasonal wind reversal.

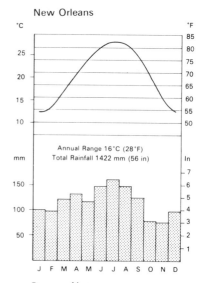

New Orleans

Annual Range 16 °C (28 °F)
Total Rainfall 1422 mm (56 in)

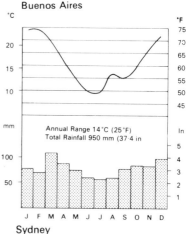

Buenos Aires

Annual Range 14 °C (25 °F)
Total Rainfall 950 mm (37.4 in)

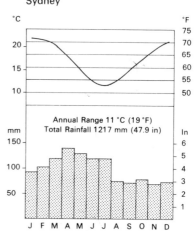

Sydney

Annual Range 11 °C (19 °F)
Total Rainfall 1217 mm (47.9 in)

4 Most of the rain falls in summer and it is convectional. The ligher rains of winter are caused
by depressions developing in the off-shore westerly winds. Total annual rainfall is about 1000 mm
(40 in).

5 Typhoons (south China) and hurricanes (south-
east U.S.A.) are common in summer.

Agricultural development

1 This climate is especially well suited to intensive
and continual crop cultivation. The temperature
and rainfall patterns enable crops to be grown
throughout the year.

2 Padi (often two crops a year) is cultivated extensively in China and it forms the basis of the
daily diet of most Chinese in central China. Padi
farming relies heavily upon human labour and
it is grown on a subsistence basis.

3 A great variety of crops are grown in south-eastern
U.S.A. Of these, cotton, maize and tobacco are
the most important. Most of the maize is used
for fattening pigs and cattle and hence animal
farming is also important.

4 In south-east Africa, especially in Natal, maize is
extensively grown as a basic food crop, and sugar
cane, tobacco and cotton are grown as cash crops.

5 Cattle and sheep are reared in vast numbers in
that part of South America which has this climate.
Some wheat and maize are grown as well.

6 In south-east Australia the main activity is dairy
farming and this region produces large quantities
of milk, butter and cheese.

COOL TEMPERATE CLIMATES

COOL TEMPERATE WESTERN MARGIN CLIMATE – THE BRITISH TYPE (Vancouver, Valencia, Hobart)

Location

1 It occurs on the western sides of continents
between 45°N. and 60°N. and south of 45°S.

2 It is best developed in north-west Europe, western
Canada (British Columbia), coastal southern
Chile, Tasmania and South Island of New
Zealand.

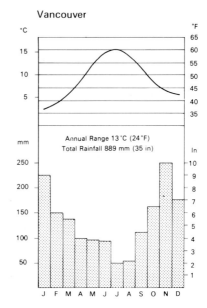

Vancouver

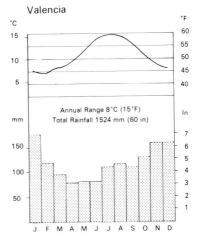

Valencia

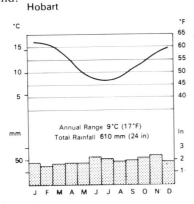

Hobart

Climatic characteristics

1 Winter temperatures range between 2°C (about
35°F) and 7°C (about 45°F), while summer temperatures range from 13°C (about 55°F) to 15°C
(59°F). The annual temperature range is between
8°C (14°F) and 11°C (24°F).

2 Prevailing winds throughout the year are from
the west, but they blow more strongly and
persistently in the winter.

3 The on-shore westerlies, especially in the Northern
Hemisphere, cross warm ocean currents (North
Atlantic and North Pacific Drifts) which prevent
winter temperatures from falling very low. It is
the combined effect of ocean currents and winds
which results in a small annual temperature
range.

4 Sub-tropical and sub-polar air meet in these
latitudes and this gives rise to depressions and

157

anticyclones which move in from the west. These pressure systems produce changeable weather.

5 Rain falls throughout the year, though there is a maximum in winter. Cyclonic and orographic rains both occur. The total annual rainfall is about 760 mm (30 in) though in mountainous regions this may be as high as 2500 mm (100 in).

Note Westerly winds and the warming influence of the warm ocean currents affects western coastal regions of Alaska and Norway which lie in the cold temperate latitudes. Winter temperatures in these regions are lower, often dropping to just below freezing point. However, these are still much higher than for places in the same latitudes and which are a few hundred miles to the east. Summer temperatures are usually below 15°C (59°F).

Agricultural development

1 A large part of the natural vegetation of deciduous forest has long since been cut down to make way for farming.
2 Extensive areas, especially in the wetter regions, are under grass and cattle and sheep farming is very important.
3 In north-west Europe the following types of farming take place:
   (i) beef cattle and dairy farming
   (ii) sheep farming for wool and meat
   (iii) cereal farming, especially for wheat, barley, and oats
   (iv) mixed farming – cattle, root crops and cereals on a crop rotation basis
   (v) market gardening, especially near to the urban conurbations
   (vi) fruit farming.
4 Agricultural activities in British Columbia are similar to those of north-west Europe except that fruit farming, especially apple farming is one of the major activities.
5 Vast flocks of sheep are reared in South Island of New Zealand and sheep farming for wool and meat dominates the economy.
6 Sheep farming is also important in Tasmania and southern Chile, but the latter region is the least developed of all regions having this type of climate. Fruit farming, especially for apples, is a major activity in Tasmania.

COOL TEMPERATE EASTERN MARGIN CLIMATE – LAURENTIAN TYPE (Montreal, Peking and Bahia Blanca)

Location

1 It occurs on the eastern sides of North America and Asia between 35°N. and 50°N. and on the eastern side of South America, south of 40°S.
2 It is best developed in the Maritime Provinces

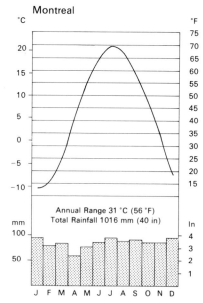

Montreal

Annual Range 31 °C (56 °F)
Total Rainfall 1016 mm (40 in)

of eastern Canada, and the New England states of the U.S.A., northern China, Manchuria, Korea and northern Japan.

Climatic characteristics

1 Winter temperatures range from −9°C (about 15°F) to −7°C (about 20°F), and summer temperatures 15°C (60°F) to 24°C (about 75°F). The annual temperature range is therefore high and it averages about 24°C (45°F). Compare this with the temperature range of 11°C (20°F) for Western Margin climates.
2 Cold winds blow seawards from the interiors of North America and Asia in the winter and this causes the temperatures to be low. In North America these winds pick up moisture from the Great Lakes which gives rise to heavy falls of snow. In Asia the out-blowing winds are dry.
3 Off-shore currents, the Labrador Current (N. America) and the Kuril Current (Asia) reduce the winter temperatures along the coasts still further. Compare with the warm currents off the western coasts in the same latitude.
4 Precipitation (as rain and snow in N. America and N. Japan) occurs throughout the year and is fairly evenly distributed. In north-east Asia (except N. Japan and N. Korea) however, rainfall is confined to the summer (see (2) above). The total annual rainfall varies from 530 mm (25 in) to 1000 mm (40 in). The rain is both convectional and cyclonic.
5 North-east Asia has a typical monsoon wind pattern, i.e. seasonal wind reversal. Dry cold out-blowing winter winds give no rain to the mainland but heavy snowfalls occur in northern Japan and northern Korea. Moist, in-blowing summer winds bring rain to all parts.

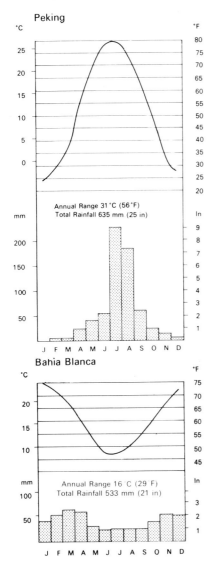

Peking

Annual Range 31 °C (56 °F)
Total Rainfall 635 mm (25 in)

Bahia Blanca

Annual Range 16 C (29 F)
Total Rainfall 533 mm (21 in)

COOL TEMPERATE INTERIOR CLIMATE
(Moscow and Winnipeg)

Location
1 It occurs in the interior of North America and Eurasia between latitudes 35°N. and 60°N.
2 It is best developed in the provinces of Alberta, Saskatchewan and Manitoba, in Canada, the north-central and mid-west of the U.S.A., in central and eastern Europe, and western U.S.S.R.

Climatic characteristics
1 Winter temperatures often fall to as low as −19°C (about 0°F) with summer temperatures rising to

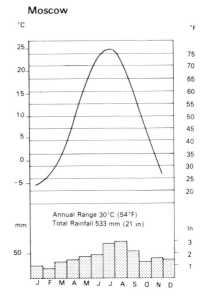

Moscow

Annual Range 30°C (54°F)
Total Rainfall 533 mm (21 in)

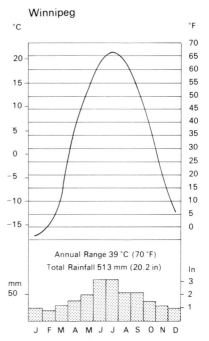

Winnipeg

Annual Range 39 °C (70 °F)
Total Rainfall 513 mm (20.2 in)

Agricultural development
1 Mixed farming with cattle, hay, oats and wheat is a major activity in the lowlands of north-east America. Dairy farming and market gardening are important near the towns. Fruit farming, especially for apples, is very important in some areas, e.g. Nova Scotia (Canada).
2 Crop farming is extensive in Manchuria and other parts of mainland north-east Asia which have this type of climate. Important crops are wheat, maize, kaoliang (a millet) and soya bean. The latter is rich in proteins and provides oil which is used in much the same way as coconut oil or olive oil. The plant also enriches the soil and is used in crop rotation.
3 Sheep farming is about the only important agricultural activity in south-eastern South America (Patagonia). The rainfall here is very low and grasslands are poor. Both mutton and wool are produced.

18°C (about 65°F) which gives an annual temperature range of 37°C (65°F).

2 Rainfall occurs mainly in the summer and it is convectional. The annual total rarely exceeds 513 mm (20.2 in). Rainfall decreases towards the west in N. America and towards the east in Eurasia. Rainfall is so low east of the Caspian Sea that the climate is of the Desert type and it is better described as Interior Desert (see the map on page 148).

Agricultural development

1 The cultivation of wheat, using large farm units, is of major importance in both the American and Eurasian regions.

2 Mixed farming, usually cattle with wheat and other temperate cereals, takes place in the European and Soviet regions.

COLD TEMPERATE CONTINENTAL CLIMATE – SIBERIAN TYPE (Verkhoyansk and Dawson)

Location

1 This is located between the Cool Interior Climate and the Tundra Climate in both N. America and Eurasia.

2 It is best developed in Canada and the U.S.S.R.

Climatic characteristics

1 Winter temperatures range from −34°C (about

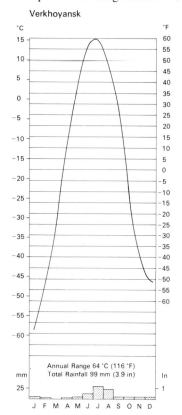

Verkhoyansk

Annual Range 64 °C (116 °F)
Total Rainfall 99 mm (3.9 in)

−30°F) in parts of Canada to −45°C (about −50°F) in parts of the U.S.S.R. Summer temperatures average about 21°C (about 70°F). The annual temperature range is normally over 55°C (100°F).

2 Total annual rainfall rarely exceeds 380 mm (15 in) and most of this occurs in the summer. These rains result from the entry of moist sea air. Snow falls in winter in eastern Canada are produced by depressions which move eastwards along the St. Lawrence Valley.

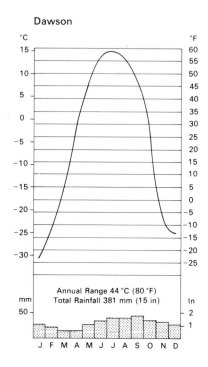

Dawson

Annual Range 44 °C (80 °F)
Total Rainfall 381 mm (15 in)

Agricultural Development

The subsoil is frozen for most of the year and this prevents most types of agriculture from taking place.

TUNDRA CLIMATE (Barrow Point and Bulun)

Location

1 It occurs in the northern continents north of the cold temperate continental climate.

2 It is best developed in northern Canada, and northern Asia.

Climatic Characteristics

1 Winter temperatures range from −29°C (about −20°F) to −40°C (−40°F) and summer temperatures are about 10°C (50°F). The annual range varies from 39°C (70°F) to 50°C (90°F).

2 Winter nights are long with hardly any daylight and summer days are long with hardly any night.

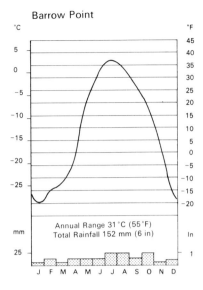

Barrow Point

Annual Range 31°C (55°F)
Total Rainfall 152 mm (6 in)

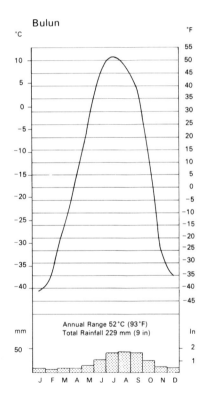

Bulun

Annual Range 52°C (93°F)
Total Rainfall 229 mm (9 in)

3 The total annual precipitation is about 250 mm (10 in) some of which falls as rain in the summer and some as snow in the winter.
4 Humidity is always low because of the low temperatures.

Agricultural development
Subsoils are permanently frozen and there is no agriculture of any type.

POLAR CLIMATE

Location
This occurs in Greenland, interior Iceland and in Antarctica.

Climatic characteristics
1 Temperatures are permanently below 0°C (32°F).
2 Blizzards are frequent.
3 Winters are really one continuous night and 'summers' one continuous day.

MOUNTAIN CLIMATE (Pike's Peak and Quito)

Location
This type of climate is best developed in regions of young fold mountains, e.g. the Rocky Mountains, the Andes, and the Himalayas, and associated mountains.

Climatic characteristics
1 In general, pressure and temperature decrease with altitude while precipitation increases. How-

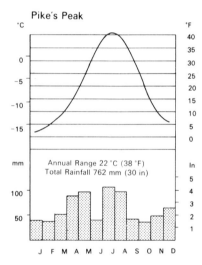

Pike's Peak

Annual Range 22°C (38°F)
Total Rainfall 762 mm (30 in)

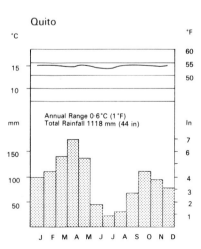

Quito

Annual Range 0·6°C (1°F)
Total Rainfall 1118 mm (44 in)

161

ever, if mountains are high enough, there is a height at which maximum precipitation occurs and above which it decreases as the moisture content of the air falls because it becomes rarefied and its temperature falls.

2 Because the air in high mountain regions is rarefied and relatively dust-free, it cannot absorb much heat. The air is therefore always cool and the daily temperature range is small. In comparison the ground absorbs much more heat in the summer days but loses it rapidly during the night. The daily ground temperature range is much greater.

3 There is usually a succession of temperature belts, not unlike those extending from the equator to the poles, in very high mountain regions. Such a succession is particularly well developed in the Andes.

4 Local winds develop in most mountain regions. These tend to blow up the valleys and up the mountain slopes in the day, whilst the winds blow down the valleys and mountain slopes in the night. Other local winds like the Chinook (in the Rockies) and the Föhn (in Switzerland) are caused by air crossing mountain ranges and descending down the lee side.

EXERCISES

1 Briefly describe the main characteristics of an Equatorial climate and its vegetation, and state what difficulties these present to agricultural development.

2 What are the main temperature and rainfall differences between Cool Temperate Western and Eastern Marginal climates?

3 Explain the following statements, using diagrams or sketch-maps, to illustrate your answers:
   (i) the characteristic aspects of a tropical monsoon climate are land and sea breezes on a continental scale
   (ii) the west coasts of continents within the tropics receive less rain than the east coasts

4 The following are the characteristics of three different types of climate:
   (a) A hot and a warm season with an annual temperature range of about $-7°C$ (about $20°F$) and with most rain falling in the hot season
   (b) Well distributed heavy rainfall with high temperatures throughout the year
   (c) A very large daily temperature range with annual rainfall usually below 120 mm (5 in)

   (i) name the type of climate for each
   (ii) choose *one* of the climates named, describe its temperature and rainfall pattern and account for this
   (iii) for *each* climatic type named, give *one* region which has that climate.

5 Explain and account for the following:
   (a) The daily temperature range of In Salah (Libyan Desert) is higher than its annual temperature range.
   (b) Hot deserts are usually located on the western sides of continents.
   (c) Rainfall decreases from 60°N. to the Tropic of Cancer on the west side of a continent and increases, in the same direction, on the east side.
   (d) Desert plants are able to survive for long periods without water.

6 **A** Altitude 60 m (197 ft): Latitude 32°

	J	F	M	A	M	J	J	A	S	O	N	D	Tot
Temperature (°C)	22	23	22	19	16	13	12	13	14	16	18	22	
(°F)	73	74	71	66	60	56	55	56	58	60	65	71	
Rainfall (mm)	8	8	17	43	124	167	162	142	83	53	20	15	84
(in)	0·3	0·3	0·7	1·7	4·9	6·6	6·4	5·6	3·3	2·1	0·8	0·6	3?

B Altitude 16 m (51 ft): Latitude 30°

	J	F	M	A	M	J	J	A	S	O	N	D	
Temperature (°C)	12	13	17	20	24	27	28	27	25	21	16	13	
(°F)	54	56	63	68	75	80	82	81	78	69	61	56	
Rainfall (mm)	114	104	114	119	106	139	167	144	116	88	93	119	14
(in)	4·5	4·1	4·5	4·7	4·2	5·5	6·6	5·7	4·6	3·5	3·7	4·7	5

Note: Fahrenheit equivalents are approximate.

Study the figures given for the climate of the two places **A** and **B**. For each of these places:
   (a) State whether it is north or south of the equator and give reasons to support your answer.
   (b) Briefly describe the characteristic features of (i) the distribution and amount of rainfall, and (ii) the temperature.
   (c) State the type of climate for each of the places **A** and **B** and name *one* region where it occurs.
   (d) Choose *one* of the climatic types named in (c)
   (e) Briefly explain why it occurs in the region that you name.

7 Each of the following statements summarises a specific type of climate: Name the climate for *each* statement and then for each of these, (i) name one region where it occurs, (ii) describe the characteristic features of its natural vegetation, and (iii) briefly describe how the climate has influenced agricultural developments.
   (a) Mild winters, 0° to 6·6°C (32° to 44°F), warm summers usually over 15·5°C (60°F); rain throughout the year being brought by westerly winds.
   (b) Warm winters, over 15·5°C (60°F), and hot summers, over 26·7°C (80°F); on-shore trade winds throughout

the year with rain falling in all months but with a maximum in summer.

(c) Mild winters, over 10°C (50°F), hot summers, over 26·7°C (80°F); light summer rainfall, mainly convectional of about 500 mm (20 in).

8 For *each* of the three towns given below (a) briefly describe the temperature and rainfall patterns, (b) state the type of climate giving reasons for your answer, (c) name one region where this type of climate occurs.

		J	F	M	A	M	J	J	A	S	O	N	D
X Altitude 676 m 2,220 ft													
Temp	(°C)	−15	−11	−5	5	11	14	16	15	10	5	−4	−10
	(°F)	5	11	23	40	51	57	62	59	50	41	24	13
Rainfall	(mm)	23	15	20	23	48	29	84	58	33	18	18	20
	(in)	0·9	0·6	0·8	0·9	1·9	3·1	3·3	2·3	1·3	0·7	0·7	0·8
Y Altitude 2,880 m 9,446 ft													
Temp	(°C)	15	15	15	14	15	14	14	15	15	15	14	15
	(°F)	59	59	59	58	59	58	58	59	59	59	58	59
Rainfall	(mm)	99	112	142	125	137	43	20	30	69	112	97	79
	(in)	3·9	4·4	5·6	6·9	5·4	1·7	0·8	1·2	2·7	4·4	3·8	3·1
Z Altitude 6 m 20ft													
Temp	(°C)	28	28	26	25	24	22	21	21	22	22	25	26
	(°F)	82	82	79	77	75	72	70	70	72	72	77	79
Rainfall	(mm)	336	376	452	399	264	282	302	203	132	99	117	262
	(in)	14·4	14·8	17·8	15·7	10·4	11·1	11·9	8·0	5·2	3·9	4·6	10·3

9 Describe and account for the temperature conditions experienced in the following types of climate: (i) warm temperate interior, (ii) cool temperate western margin, (iii) tropical continental, (iv) mountain types of climate as illustrated by the figures given below

mean monthly temperatures

Place	Altitude	Lowest	Highest
(i) **Kimberley**	1.197 m	July 10°C	January 25°C
(29°S 25°E)	3,927 ft	(50°F)	(77°F)
(ii) **Victoria**		January 4°C	July 16°C
(48°N 123°W)		(39°F)	(60°F)
(iii) **Kayes**	30 m	January 25°C	May 34°C
(14°N 12°W)	198 ft	(77°F)	(94°F)
(iv) **Uspallata**	2,845 m	June 1°C	December 12°C
(33°S 70°W)	9,335 ft	(34°F)	(53°F)

Objective Exercises

1 A Tropical Continental (savana) climate is one which has

A a hot dry summer and a wet rainy winter

B a hot summer, cooler winter with most rain falling in the summer

C high temperatures and low rainfall throughout the year

D westerly winds in the winter and trade winds in the summer

E warm, wet summer and a cool, dry winter

A B C D E
☐ ☐ ☐ ☐ ☐

2 Which one of the following stations is situated in a region which has a tropical monsoon climate?

A Durban (South Africa)

B Valencia (Ireland)

C In Salah (North Africa)

D Lagos (Nigeria)

E Bombay (India)

A B C D E
☐ ☐ ☐ ☐ ☐

3 A Mediterranean Climate is characterised by very dry summers because

A relative humidity is low

B skies are cloudless

C trade winds blow from the land to the sea

D there are no on-shore winds

E temperatures are high

A B C D E
☐ ☐ ☐ ☐ ☐

4 This is a description of a specific climatic type "Warm winter, over 15.5°C, and hot summers, over 26°C; on-shore trade winds blow throughout the year bringing rain in all months but with a maximum in summer." Which one of the following climatic types does this describe?

A Warm Temperate Western Margin

B Tropical Monsoon

C Warm Temperate Eastern Margin

D Tropical Marine

E Equatorial

A B C D E
☐ ☐ ☐ ☐ ☐

5 The climate of Singapore differs from that of a station on the Mediterranean coast during July in that it experiences

A higher average temperatures and a higher average rainfall

B higher average temperatures and a lower average rainfall

C lower average temperatures and a higher average rainfall

D lower average temperatures and a lower average rainfall

A B C D
☐ ☐ ☐ ☐

6 There are very few low-growing plants in a tropical humid forest because there is

A little sunlight at ground level because of the canopy of tall trees

B lack of moisture at ground level because of the covering of tall trees

C a thick covering of leaves and decaying vegetation on the forest floor which prevents plant growth

D a high evaporation rate on the forest floor which causes plants to wither

A B C D
☐ ☐ ☐ ☐

7 Which of the following statements best explains the difference between the climate of The British Isles and that of Southern Portugal?

A The British Isles are nearer the North Pole.

B The British Isles lie under westerly winds for most of the year.

C The western parts of the British Isles are mountainous.

D The British Isles are surrounded by water.

A B C D
☐ ☐ ☐ ☐

8 What type of climate occurs on the western side of continents between latitudes 30°N and 45°N?

A British Type

B China Type

C Mediterranean

D Hot Desert

A B C D
☐ ☐ ☐ ☐

9 All of the following types of climate occur in the Southern Hemisphere **except**

A Hot Desert

B Savana

C Equatorial

D Tundra

A B C D
☐ ☐ ☐ ☐

10 The climate of Western Europe is mild for its latitude because

A the influence of altitude on temperature is minimal

B the prevailing winds are the Westerlies

C the coast is washed by the North Atlantic Drift

D the region is within the Low Pressure belt

A B C D
☐ ☐ ☐ ☐

11 Which of the following tropical climates has the largest diurnal temperature range?

A Tropical Monsoon

B Equatorial

C Hot Desert

D Savana

A B C D
☐ ☐ ☐ ☐

12 Which of the following statements best describes the climate represented by the graph given below?

A The annual temperature range is fairly high and rainfall occurs mainly in the winter.

B The summer is hot and dry.

C The summer is hot and dry and the winter is mild and rainy.

D Most of the rain falls in the winter when temperatures are below 15°C

A B C D
☐ ☐ ☐ ☐

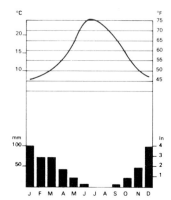

15   Vegetation

Geographers divide the world's vegetation into a number of types according to the appearance of the plants. The basic types, shown in the map on page 169 are:

1 Forest 2 Grassland 3 Desert; but before examining these it is important for us to know something about what a plant requires in order to maintain its growth.

PLANT REQUIREMENTS

Green plants make their own food by using:
(i) Water
(ii) Sunlight
(iii) Carbon dioxide
(iv) Mineral salts.

Water with the mineral salts in solution enters the roots of a plant from the soil and passes through the stem into the leaves, where a process called *photosynthesis* goes on. In this process carbon dioxide, which enters the leaves from the air, combines with the water in the presence of sunlight to form carbohydrates (food). The surplus water passes out of the leaves into the air and this movement of water is called *transpiration*. These two processes enable plant growth to take place. If any of the four requirements are missing or are inadequate then these processes either slow down or cease altogether.

The Influence of Temperature and Water on Plant growth

Temperature and water are the two most critical plant requirements because they do not occur in adequate amounts in all regions, and in some they vary in amount from season to season.

Plant growth normally ceases when the temperature falls below 6°C (about 42°F). In polar regions the temperature is always below 6°C (about 42°F) and there is no plant life. In other regions the temperature is always above 6°C (above 42°F) and continuous plant growth is possible. In still other regions the temperature is below 6°C (about 42°F) for one season and above it for another season. Plant growth is therefore seasonal.

Plants, like animals, have become adapted to their physical environment, in particular to the water and temperature aspects of this. Because water and temperature vary from natural region to natural region, the plants also vary in size and form. Thus hot wet regions normally have a forest vegetation whereas regions having light seasonal rains usually support a grass vegetation. Generally speaking, trees require more water than do grasses

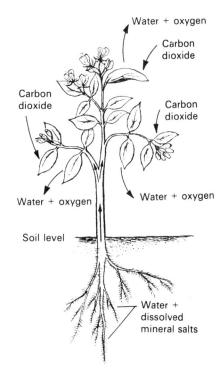

During sunlight

Water + oxygen
Carbon dioxide
Carbon dioxide
Carbon dioxide
Water + oxygen
Water + oxygen
Soil level
Water + dissolved mineral salts

which in turn require more than desert plants. We need not examine all the ways in which plants have adapted themselves to their environments, but we must have a look at the influence of drought and cold on plant adaptation.

Influence of Drought on Plants

1 Some plants develop long roots to reach water supplies far below the surface.
2 Some plants develop water storage organs, e.g. the baobab tree stores water in its trunk.
3 Some plants have special leaves which reduce transpiration, e.g. thorn-like leaves, rolled-up leaves, leaves with waxy surfaces.
4 Some plants shed most of their leaves when the dry season is also hot. If the plants do not do this then they would literally dry up as transpiration proceeds. Monsoon forests are almost leafless in the dry season.

Influence of Cold on Plants

When temperatures fall below 6°C (about 42°F) for several months of the year, many plants are unable to obtain sufficient water from the soil and many

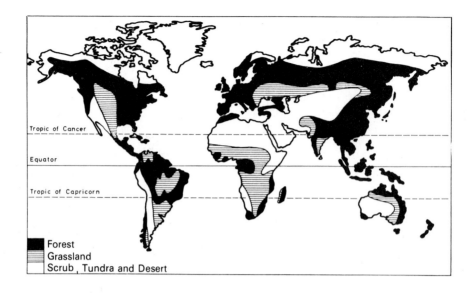

of them shed their leaves. Such plants are said to be *deciduous*. This reaction to cold is particularly common in temperate forests.

Note Some trees have become adapted to cold weather and do not shed their leaves. *Coniferous trees* (except larch) keep their leaves throughout the cold season and they are called *evergreen* trees. They are able to do this because:

(i) their leaves are rolled and little transpiration takes place from them.

(ii) they need less water than other trees.

TYPES OF NATURAL VEGETATION

The map on page 169 shows the basic types of natural vegetation. This map gives a very simplified picture of the vegetation pattern, and so does the map on page 170, although this map shows more detail. It would appear from these maps that natural vegetation covers the entire land surfaces of the earth, but, of course, this is not so. Extensive low-land regions have long since been cleared of their vegetation to make way for man's crops and settlements. These maps are intended to show what the natural vegetation pattern would be like if man and his animals had in no way interfered with the land. Natural vegetation is only extensive today in those parts of the world where man finds great difficulty in mastering the environment, e.g. the Arctic lands and the equatorial river basins.

We will now examine these basic vegetation types.

For each type we must know the location and the names of some of the more common plants together with the way in which these plants have adapted themselves to the climatic characteristics.

THE FORESTS

I Tropical Evergreen Forest

Location   Amazon and Zaire Basins; West African coastlands; Malaysia; coastal Burma, Cambodia and Vietnam; most of Indonesia and New Guinea.

Characteristics   1 Contains a great variety of plants which are close together.

2 The forest consists of three layers:

(a) *top layer*: tall trees with buttress roots;

(b) *middle layer*: tree ferns, lianas, e.g. rattan, and epiphytes, e.g. orchids;

(c) *bottom layer*: ferns, herbaceous plants and saprophytes.

3 Nearly all the trees are broad-leaved evergreens because high temperatures and evenly distributed rainfall permit growth throughout the year.

4 Absence of seasonal climatic change results in some plants being

Equatorial Evergreen Forest in Ghana

in flower, others in fruit and others in leaf-fall at one and the same time.

5 The leaves of the tall trees form an almost continuous canopy which shuts out most of the light at ground level. There is therefore little undergrowth.

6 Mangrove trees, with stilt roots, form dense forests in coastal swamps and the lower valleys of tidal rivers.

7 When a part of a tropical evergreen forest is cleared, either for shifting cultivation or for lumbering, a less luxuriant forest growth takes over. This is called *secondary forest* and it consists of short trees and dense undergrowth. In Malaysia it is known as *belukar*.

Examples of trees *Mahogany*, *ebony*, *rosewood*, *ironwood* and *greenheart* are common trees. *Palms* and *tree ferns* also occur in most equatorial forests.

These trees, as well as most of the other plants, occur singly or in small groups. There are rarely extensive stands of a particular tree.

II Tropical Monsoon Forest

Location Burma; Thailand; Cambodia; Laos; N. Vietnam; parts of India; east Java and the islands to the east; N. Australia.

Characteristics 1 There is a smaller number of species than in the tropical evergreen forest.

2 Most of the trees are deciduous, losing their leaves in the hot, dry season which is a period of rest for them.

3 Heavy rain and high temperatures in the wet season result in rapid growth and the trees soon become covered with leaves.

4 The trees are tall, often as high as 30 metres (approx. 100 feet), but they are not as close together

167

Mangrove Swamp

as they are in tropical evergreen forests. Because of this, undergrowth is more dense (more light reaches the ground). Bamboo thickets are common.

5 Most monsoon forests contain valuable hardwoods, e.g. teak.

6 These forests merge into equatorial forests in regions where the dry season becomes short or non-existent, and into grasslands in regions where the wet season becomes short and rainfall less heavy.

Examples of trees

Teak (especially in Burma, Thailand, Cambodia, Laos and E. Java); *bamboo* (especially in Thailand, Cambodia, Laos and Vietnam); *sal*; *sandalwood*; and *lianas. acacia, eucalyptus* and *casuarina* occur in E. Java and N. Australia.

Tropical Monsoon Forest

Natural Vegetation – the World

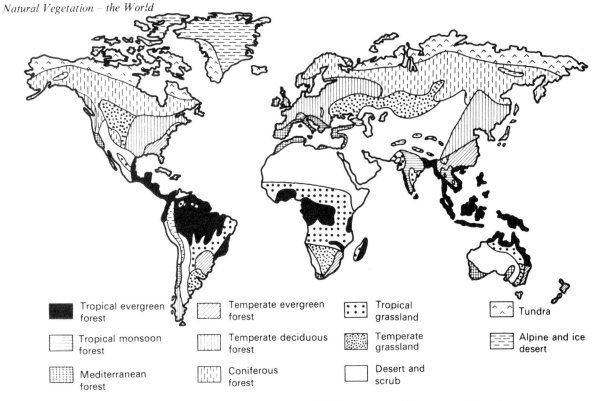

Legend:

- ■ Tropical evergreen forest
- Tropical monsoon forest
- Mediterranean forest
- Temperate evergreen forest
- Temperate deciduous forest
- Coniferous forest
- Tropical grassland
- Temperate grassland
- Desert and scrub
- Tundra
- Alpine and ice desert

Structure of an Equatorial Forest

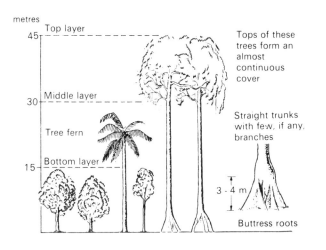

metres — Top layer 45 — Middle layer 30 — Tree fern — Bottom layer 15

Tops of these trees form an almost continuous cover

Straight trunks with few, if any, branches

3 - 4 m

Buttress roots

The diagram shows the three layers of vegetation in an equatorial forest. Many of the tall trees have buttress roots which give them support (remember they are of great height). Trees with buttress roots also occur in monsoon forests.

III Temperate Evergreen Forest

Location

This occurs chiefly on the eastern sides of land masses in the warm temperate latitudes: S. China; S. Japan S.E. Australia; North Island (New Zealand); Natal coastlands (Africa); S. Brazil, and S.E. states of the U.S.A.

Characteristics

1 Most of these regions have rain throughout the year with winter temperatures often over 10°C (50°F) which means that plant growth can go on all the year. Most of the trees are broadleaved evergreens, although there are deciduous trees as well, especially in the Gulf States of the U.S.A.

2 Most of the evergreens are hardwoods e.g. quebracho (S.E. Brazil) and cedar.

3 The forests of China and southern Japan contain evergreen oak, magnolia and camphor trees, all of which are of economic value.

4 Walnut, oak, hickory and magnolia trees form the bulk of the forests in the Gulf States of the U.S.A.

5 The highlands of south-east

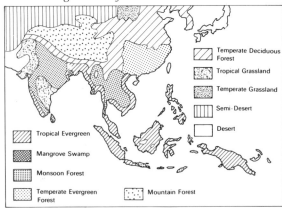

Natural Vegetation of South-East Asia

Tropical Evergreen

Mangrove Swamp

Monsoon Forest

Temperate Evergreen Forest

Mountain Forest

Temperate Deciduous Forest

Tropical Grassland

Temperate Grassland

Semi-Desert

Desert

Examples of trees

Africa, especially in Natal, contain ironwood, blackwood and chestnut trees, as well as wattle trees whose trunks are used as pit props in the coal mines of Natal.

6 The forests of south-east Australia are famous for eucalyptus trees while those of the mountain slopes of south-east Brazil are equally well known for the Parana pine trees.

7 Most of these forests look very much like tropical forests in that the vegetation is thick and profuse. *Evergreen oak*; *magnolia* (especially China and the U.S.A.); *camphor* and *bamboo* (especially China);

Temperate Evergreen Forest in North Island (New Zealand)

cedar (evergreen), *maple* and *walnut* (both deciduous); *eucalypts* (Australia); *tree fern* (New Zealand); *mulberry* (China); and *cypress* (U.S.A.).

IV Mediterranean

Location This occurs chiefly on the western sides of land masses in the warm temperate latitudes. Lowlands around the Mediterranean Sea; S.W. Australia and the Adelaide District of Australia: S.W. Africa; central Chile; central California.

Characteristics 1 Originally the vegetation was forest but much of it has been cut down and the rest has been ravaged by foraging goats so that all that now remains is a scattered woodland-type of vegetation.

2 Summers are hot and dry which make plant growth difficult. However, the plants of this region have become adapted to the summer drought by storing water obtained from the winter rains, in leaves and bark. Many plants have waxy, spiny or small leaves to cut down the rate of transpiration. Other plants, e.g. grape vine, have long tap roots which can reach down to moist rock layers well below the surface.

3 In the drier parts the vegetation becomes scrub-like and consists of sweet-smelling herbs and shrubs such as lavender, rosemary, thyme and oleander.

4 In the wetter parts, e.g. on mountain slopes, coniferous trees are common.

Examples of trees *Evergreen oak*; *cork oak*; *eucalypts*, *jarrah* and *karri* (S.W. Australia); *cedar*, *cypress*, and *sequoia* or redwood (California), all of which are conifers. The scrub vegetation which includes *lavender*, *rosemary*, *myrtle* and *oleander* is a secondary type of vegetation, i.e. it has arisen in consequence of the destruction of the original vegetative cover. In France this is called *maquis*; and in California *chaparral*.

V Cool Temperate

Location W. and central Europe; eastern U.S.A.; N. China; N. Japan;

Mediterranean Vegetation in Greece

Korea; S. Chile and South Island (New Zealand). The forest is poorly developed in S. Chile because of the high relief.

Characteristics 1 In most of these regions, winter temperatures fall below 6·1°C (43°F), the minimum temperature for plant growth, which results in most of the trees shedding their

Temperate Deciduous Forest

leaves and becoming deciduous.

2 Some of the deciduous trees are hardwood and the most common are, oak, beech, birch, hornbeam and ash.

3 Many of the trees occur in pure stands and they are of great economic value.

Examples of trees   Oak; *beech*; *hazel*; *elm*; *chestnut*; *poplar*, and, in N. America, *walnut*; *maple*; *hickory*; *cedar* and *spruce* (both conifers).

VI Coniferous Forest

Location   This is most extensive in high latitudes and on high mountains, although it does develop on sandy soils in warmer regions. There are two main belts of this forest: (i) across Eurasia extending from the Atlantic to the Pacific, (ii) across N. America extending from coast to coast.

Characteristics   1 Most of the trees are evergreen and they are coniferous. They are especially well adapted to the long, cold and often snowy winters, by growing needle-shaped leaves which reduce transpiration to a minimum. Further, the leaves have a tough thick skin which protects them from winter cold. The tree is conical and has flexible branches which allow the snow to slide off. It has widely spread shallow roots to collect water from the topsoil, above the permafrost layer.

Deciduous tree

Tree in leaf

Same tree after leaf fall

Deciduous broad leaf

Coniferous tree

Tree as it is all the year

Pine cone

Coniferous needle leaf

Temperate Coniferous Forest

2 Coniferous trees produce soft wood which is in great demand for making paper, especially newsprint, matches, furniture and synthetic fibres such as rayon.

3 These forests have practically no undergrowth because the soil is frozen for many months each year.

Examples of trees

Hemlock, spruce, pine and fir. The trees of coniferous forests in Mediterranean regions are chiefly *cypress* and *cedar*.

THE GRASSLANDS

I Tropical Grassland

Location

These are located mainly in the continental regions of tropical latitudes, where the rains occur in the hot season which lasts for about 5 months; north and south of the Zaire Basin, West Africa and the East African Plateaus; parts of Brazil; the Guiana Highlands; north and east of the Australian Desert; and parts of the Indian Deccan Plateau.

Characteristics

1 Although tall grasses form the dominant plant life of these regions, trees are common, especially near water courses and in the more humid areas, that is, where the region merges with equatorial forest.

2 The grasses are usually 2 metres (approx. 6 feet) high, or more, and they grow in compact tufts. Their long roots can reach down to the moist rock layers far below the surface. During the dry season the leaves of the grasses turn yellow and die but the roots remain dormant (sleeping). These grasses are therefore deciduous.

3 The trees of these regions are also deciduous, shedding their leaves in the dry season. Some

Baobab tree

Temperate Grassland in South Africa

trees, e.g. baobabs and bottle trees, store water in their swollen trunks and in this way they are able to survive the dry season.

4 Where these grasslands merge with the hot deserts, the vegetation changes greatly. There is no longer a continuous vegetation cover: instead, clumps of scrub-like plants scatter the surface. *Mallee*₁ and *mulga*₂ characterise this type of vegetation in Australia.

Grassland names

Tropical grasslands have different names according to their location, e.g. *Campos* (Brazil); *Llanos* (Guiana Highlands); *Savana* (Africa and Australia).

1 *Mallee* – eucalyptus bushes set in a thicket of
– coarse grass.
2 *Mulga* – clumps of acacia set in a thicket of coarse
– grass.

II Temperate Grassland

Location

These are best developed in the continental interiors of temperate latitudes, e.g. hearts of Asia and N. America. Less extensive areas occur in S. Africa, S. America and Australia.

Characteristics

1 These grasslands are almost treeless and they contrast sharply with the tropical grasslands.

2 In the moister regions (rainfall over 500 mm or 20 in) the grasses are tall, though not as tall as those of tropical grasslands, and they are nutritious. These grasses are typical of the Black Earth region of the Ukraine (southern U.S.S.R.) and the moister parts of the American Prairies (now mainly under wheat).

In the drier regions (rainfall under 500 mm or 20 in), the grasses are shorter, tougher and less nutritious. These grasses are typical of the High Plains (U.S.A.) and the Asian Steppes.

3 During the heat of summer the grass begins to wither and most of it dies in the autumn. The roots do not die, and with the coming of spring, new leaves form and the land is once again covered with a green mantle.

4 A few trees, such as poplars, willows and alders, grow in the damper soils flanking water courses.

5 On the poleward side, the temperate grasslands merge with the coniferous forests while on the equatorward side, they merge with scrub of semi-deserts.

Grassland names The following names are generally recognised: *steppe* (Eurasia); *prairie* (N. America); *pampas* (Argentina); *veldt* (S. Africa); *downs* (Australia).

Note In most of the grasslands the leaves of the grasses wither and die in the dry season (tropics) or cold season (temperate regions). The roots of the plants however do not die, and, in the following season when the rains come, the aerial parts of the plants grow afresh.

DESERT AND SEMI-DESERT

I Tropical Desert

Location Usually lies between 15°N. and 30°N. and 15°S. and 30°S. They lie on the western sides of land masses except for Africa where they extend

Giant Cactus from the Sonora Desert – Mexican Border with the U.S.A.

from coast to coast, linking up with the Asian deserts. The chief regions having a desert vegetation are: Sahara (N. Africa); Arabia; parts of Iran, Iraq, Syria, Jordan and Israel; part of Pakistan; central Australia; the Namib Desert (S.W. Africa); Atacama (coastal Peru and N. Chile); and S. California, N. Mexico, parts of Arizona (N. America).

Characteristics 1 Only very small parts of tropical deserts are without any type of vegetation.

2 Desert plants are all very special in that they can withstand high temperatures and long periods when no rain at all falls.

3 The plants have become adapted to conditions of extreme drought in several ways. Some plants have long roots, others have few or no leaves, and what there are are tough, waxy or needle-shaped to

reduce transpiration to a minimum.
4 Many plants produce seeds which lie dormant for years until a little rain falls and then they germinate.
5 The most common plants are cacti, thorn bushes and coarse grasses.

II Semi-desert and Scrub

Location   This type of vegetation occurs in regions which either border the tropical deserts, or in the interiors of continents where the rainfall is just insufficient to maintain a continuous cover of vegetation.

Characteristics   In tropical scrubland the chief plant is the thorny acacia. During the period of rains a short-lived but rich mantle of grasses and flowering plants covers extensive areas. Sage bush is common in N. American scrubland. Mulga is sometimes included in this vegetation type.

III Tundra

Location   This type of vegetation is chiefly confined to the Northern Hemisphere, fringing the Arctic Ocean in the continents of Eurasia and N. America.

Characteristics   1 The growing season is very short (about two months), and at this time the surface soil thaws but the subsoil remains frozen. Water therefore lies on the surface. This influences the pattern of vegetation.
2 Where the tundra meets the coniferous forest there are stunted willows and birches.
3 Most of the tundra consists of a low cover of vegetation made up of lichens, mosses, sedges, and flowering shrubs such as bilberry and bearberry.

EXERCISES

1 Carefully describe the influences that high temperatures, low temperatures and drought can have on plants, and by choosing specific examples, show how plants have become adapted to hot, dry conditions.
2 Make a list of the main forest types and for *two* of these:
  (i) name *two* types of common tree
  (ii) state whether the trees are deciduous or evergreen

  (iii) name *one* region where this type of forest occurs.
3 Carefully explain, by using appropriate diagrams, where relevant, the meanings of the following statements:
  (i) the hearts of continents have few trees
  (ii) the undergrowth of equatorial forests is not as well developed as is that of tropical monsoon forests
  (iii) vegetational zoning takes place both latitudinally and vertically.
4 Make a list of the main types of grassland and for each type:
  (i) name *two* regions where it occurs
  (ii) state any regional names it may have
  (iii) briefly describe how it is used by man.

Objective Exercises

1 All of the following are essential to the proper growth of green plants **except**
A water
B sunlight
C carbon dioxide
D high temperatures
E high humidity

                    A B C D E
                    ☐ ☐ ☐ ☐ ☐

2 A forest whose broad-leafed evergreen trees grow to great heights and which are festooned with parasitic plants and epiphytes, and beneath which the undergrowth is very sparse, is known as
A Mediterranean Forest
B Temperate Evergreen Forest
C Equatorial Forest
D Coniferous Forest
E Temperate Mixed Forest

                    A B C D E
                    ☐ ☐ ☐ ☐ ☐

3 Plants which have developed special water storage organs or which can survive long periods of hot dry weather are common to regions which have
A an Equatorial Climate
B a Tropical Continental (Savana) Climate
C a Tundra Climate
D a Warm Temperate Interior Climate
E a Temperate West Margin Climate

                    A B C D E
                    ☐ ☐ ☐ ☐ ☐

4 The temperate grasslands of the interior of Eurasia are called
A prairie
B steppe
C veldt
D downs
E pampas

                    A B C D E
                    ☐ ☐ ☐ ☐ ☐

5 A specific type of natural vegetation is known by all of the following names **except**
A veldt
B steppe
C campos
D downs
E prairie

    A   **B**   **C**   **D**   **E**
    ☐   ☐   ☐   ☐   ☐

6 Some plants are in flower, others are in fruit, and others lose their leaves, all at the same time, in regions which have no seasonal climate changes. This type of plant occurs in
A Coniferous Forests
B Mediterranean Forests
C Tropical Rain Forests
D Temperate Evergreen Forests

    A   **B**   **C**   **D**
    ☐   ☐   ☐   ☐

7 A region which has a long cold winter and a short cool summer, and whose sub-soils are frozen for most of the year will have a natural vegetation of
A maquis and low scrub
B acacia and sal
C lichen and sedge
D sage bush and cacti

    A   **B**   **C**   **D**
    ☐   ☐   ☐   ☐

8 An equatorial forest may contain all of the following trees **except**
A ebony
B iron wood
C baobab
D greenheart

    A   **B**   **C**   **D**
    ☐   ☐   ☐   ☐

9 The trees of tropical rain forests and coniferous forests are similar in that
A they have large broad leaves
B their branches slope downwards
C they lose their leaves at the same time
D they are evergreen

    A   **B**   **C**   **D**
    ☐   ☐   ☐   ☐

10 What type of vegetation occurs where the following conditions prevail? (a) podzol soils, (b) annual temperature range 38°C, (c) annual precipitation about 300 mm which is fairly evenly distributed through the year.
A thorn scrub
B coniferous forest
C temperate grassland
D monsoon forest

    A   **B**   **C**   **D**
    ☐   ☐   ☐   ☐

11 Evergreen Mediterranean forest occurs
A on the eastern sides of land masses which have a warm temperate climate
B in hot wet equatorial lowlands
C on the western margin of Europe where the climate is of the cool temperate type
D between 30° and 40° N and S on the west sides of continents

    A   **B**   **C**   **D**
    ☐   ☐   ☐   ☐

12 Which of the following is associated with a Savana Climate?
A shallow-rooted trees
B trees that grow throughout the year
C scrub
D dense evergreen forest

    A   **B**   **C**   **D**
    ☐   ☐   ☐   ☐

16 Soils

GENERAL

Weathering processes break up the surface of a rock into small particles. Air and water enter the spaces between these and chemical changes take place which result in the production of chemical substances. Bacteria and plant life soon make their appearance. When the plants die, they decay and produce *humus* which is all-important to soil fertility. Broadly speaking, humus consists of the decayed remains of both plants and animals. Bacteria play a vital part in the decomposition of these remains. The end product of these chemical and biological processes is *soil*. From this it will be clear that the nature of any soil is influenced by:

(i) weathering
(ii) vegetation
(iii) parent rock
(iv) climate (this determines both the type of weathering and the natural vegetation).

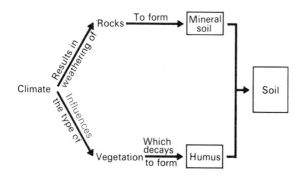

Contents of the Soil

All soils contain:

(i) mineral particles (inorganic material)
(ii) humus (organic material)
(iii) water
(iv) air
(v) living organisms, especially bacteria.

The actual amounts of each of these depend upon the type of soil. Many soils are deficient in one or more of these.

SOIL PROFILE

A soil profile is a vertical section through the soil to the underlying solid rock. Most soil profiles consist of three layers which are called *horizons*. These are lettered A, B and C.

Horizon A – the soil proper
Horizon B – the subsoil
Horizon C – the solid rock.

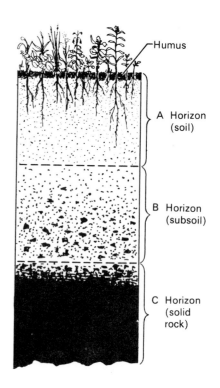

You will see from the diagram above that the top part of the A horizon is often rich in humus, and that the texture of the soil becomes coarser as the C horizon is approached.

Factors Influencing the Soil Profile

Water Movement in the Soil

When rainfall is greater than evaporation, water moves downwards in the soil and mineral matter is removed from the top layer (A horizon) and is deposited in the layer beneath (B horizon). Very often these deposits give rise to a hard layer which is called a *hardpan* and this can cause drainage to be poor. When this takes place in a soil the soil is said to be *leached*. In cold wet regions leaching helps to produce a grey soil which is known as a *podzol*; in hot wet regions it may produce a red-brown soil which is known as *laterite*.

Laterite

In humid tropical regions soil water contains very little organic matter, and such water does not dissolve iron and aluminium hydroxides. Most other minerals however dissolve and these are carried in solution to the B horizon where they are deposited. Ultimately a soil may be formed which is composed mainly of *iron* and *aluminium compounds*. This is called *laterite*. This is usually red in colour and becomes extremely sticky when wet. It is a useful material for making bricks since it sets hard on drying. Because it is the end product in the process of weathering, it is almost completely resistant to further weathering and buildings made of it last for a very long time. Some laterites are very rich in aluminium compounds and are called *bauxite*. Bauxite deposits are usually white or grey in colour. Hardpans also occur in laterites.

Note Laterite can form from any type of rock.

In hot desert and semi-desert regions there is an upward movement of water in the soil. This results in the deposition of mineral matter in the A horizon. Some of the important deposits of saltpetre have been formed in this manner.

CLASSIFICATION OF SOILS

Climate plays the greatest part in the formation of a soil (left diagram, page 178), and any classification of soils therefore tends to have a climatological basis. This means that each major soil type corresponds with a specific type of climate. These types are called *zonal soils*. They generally occur in extensive belts and they are mature soils.

Zonal Soils

A *Tropical*
   (i) Laterite
   (ii) Red Soils
   (iii) Black Soils
   (iv) Desert Soils.
B *Temperate*
   (i) Podzols
   (ii) Chernozems (Black Soils)
   (iii) Brown Soils
   (iv) Desert Soils.

For laterite, tropical red soils and tropical desert soils see notes under laterite.

Tropical Black Soils

This type of soil develops in humid tropical regions which have volcanic rocks. The soil is rich in calcium carbonate and other minerals, and when wet it becomes very sticky. It occurs extensively in N.W. Deccan (where it is cultivated with cotton), and in smaller areas in N. Argentina, Kenya and Morocco. In the Deccan it is called *regur*.

Chernozems

These are probably the richest soils in the world. They form under a natural vegetation of grass in the temperate latitudes and are rich in humus. This is because there is very little leaching. Most of the world's wheat is grown on these soils.

Temperate Brown Soils

These soils formed in climatic regions which had a natural vegetation of deciduous forest. The soils are rich in humus, though since much of the forest has now been removed for agriculture, manures have to be applied to the soils to maintain the organic content.

Temperate Desert Soils

The soils of arid temperate regions have not been leached. They are therefore rich in plant foods, and, if irrigated, they are often very fertile. The soils are grey to brown in colour and they occur chiefly in the U.S.S.R. (between the Caspian Sea and Lake Aral), and in parts of west U.S.A.

Other Soils

There are other soils which are less influenced by climate than the zonal soils. These cannot be discussed in any detail here, but the names of some of the more important will be given.

Peat Soils: they form where drainage is very poor, e.g. swamps. Vegetation does not properly decay.

Alluvial Soils: they consist of a mixture of clay, sand and silt which has been deposited by water. They form some of Asia's most important agricultural regions.

Terra Rossa Soils: These are red soils which form in limestone regions under semi-arid conditions.

Wind-deposited Soils: the chief of these is *loess*.

SOIL EROSION

Soils and Agriculture

Farmers are not generally interested in soil profiles and soil classification such as we have just discussed. What chiefly interests them is whether the soil is fertile or not. In many regions farmers have learnt how to keep the soil fertile, or to make it more fertile by adding manures, or growing different crops in rotation. But this practice is by no means universal. Soils will deteriorate if they are not properly cultivated, and soil deterioration usually results in the removal of soil by wind or water. When this happens soil erosion is said to take place.

Causes of Soil Erosion

Soil erosion is frequently caused by the action of Man and his domestic animals. Most of it begins with the removal of the natural vegetation, be it grass or forest. So long as the vegetation cover remains, there can be little if any erosion because the roots of the plants bind the soil particles together, and the vegetation itself protects the soil from the action of wind and rain. The destruction of the natural vegetation may be caused by turning the land into crop land or by putting too many domestic animals on it which results in over-grazing, or by setting fire to it which is still a common practice of shifting cultivators.

Types of Soil Erosion
I By Water

(i) *Sheet Erosion*

This affects large areas and it occurs when rain falls on a gentle slope which is bare of vegetation. This type of erosion results in the removal of a uniform depth of soil. The middle photograph shows the nature of sheet erosion in a woodland. You will see that the land is by no means bare of vegetation in this example.

(ii) *Gully Erosion*

This is more localised and occurs when heavy rainfall rushes down a steep slope, cutting deep grooves into the land. The grooves become deepened and widened to form *gullies* which finally cut up the land to give 'badlands'. This type of erosion is especially frequent in semi-arid regions.

II By Wind

Regions having a low rainfall or a definite dry season are liable to have their soils reduced to dust and blown away by the wind if the land is bare of vegetation. We have already seen how the wind

The photograph below shows the force of soil erosion. Some idea of the depth to which erosion has taken place can be obtained from the height of the trees.

Severe Soil Erosion at Ceara, Brazil

Sheet Erosion in South Africa

Gully Erosion in Kenya

Part of the 'Dust Bowl' in the U.S.A.

removes fine particles of material in desert regions by the process of deflation. The same process also take place in the marginal zones of some of the temperate grasslands which have had the grass vegetation replaced by crops such as wheat, or which have been over-grazed by cattle. In these marginal zones rainfall is not only lower than in the grasslands proper, but is also less reliable. Sufficient rain may fall for a few years to enable the crops to thrive but this is inevitably followed by a series of drier years when the crops fail and the land is abandoned. It is at this time, when the land is bare, that wind erosion sets in. One of the worst cases of such erosion occurred in the 1930s in the mid-western States of the U.S.A. The area devastated here is called the 'Dust Bowl' (see (photograph).

Common Farming Practices which Lead to Soil Erosion

1 The cultivation of crops in regions which do not have a reliable rainfall, i.e. dry years following wet years, and which have only just sufficient rainfall for crop farming.

2 The ploughing of land up and down the slope. This provides man-made channels which can be enlarged into gullies by surface run-off.

3 In shifting cultivation (top photograph, page 182) a piece of forest is destroyed by fire, and crops are grown in the soils of the cleared patch which are now enriched by wood ash. After one or two years of such cultivation, the patch is abandoned and a new clearing is made. The abandoned patch soon experiences soil erosion by rainfall. Shifting cultivation usually takes place in the wet tropics. It is a common method of crop farming with wandering forest peoples.

4 The cultivation of the same type of crop on a piece of land year after year, leads to soil depletion if manures or fertilisers are not used. Most plants usually make a demand on a particular mineral compound and if the same type of plant is grown over a number of years then the soil will become deficient in this mineral. When this happens the soil deteriorates and soil erosion may set in.

5 The cutting down of forests, especially on the higher slopes, may result in soil erosion and the spreading out of the transported soil over the lowlands where farm land can be seriously damaged. (The middle photograph on page 182 shows the sort of damage that occurs when the transported soils are deposited in built-up areas.)

Shifting Cultivation in Chile

Deposition of soil after heavy rains in Hongkong in 1958

Terrace Cultivation in N. China. The low embankments around the fields protect the soil from erosion by preventing the water from rushing down the slopes

Regions where Soil Erosion takes place
Wind Erosion
Mid-western States of the U.S.A. The eroded region is called the *Dust Bowl.*

Water Erosion
The Mediterranean countries. It is caused by over-grazing by the goat especially in the Eastern countries. Destruction of forests in Spain and parts of Italy has caused erosion.
Parts of India, Sri Lanka, Sumatra, Java and Thailand. Caused mainly by shifting cultivation.
The savana regions of Africa. Caused by shifting cultivation and over-grazing.

SOIL CONSERVATION
Soil erosion has made millions of acres of land unproductive. As the world's population increases year by year, so more and more food has to be produced if famine and disease are to be eliminated, and if all people are to get an adequate and balanced diet The governments of most countries have long since realised that soil erosion is one of Man's greatest enemies, and measures are now being taken to reduce erosion to a minimum and to reclaim land which has already been eroded. International organisations like the Food and Agriculture Organization (F.A.O.) have for many years been carrying out a programme of soil conservation. This has involved the introduction of sound methods of farming which not only prevent erosion but which also keep the soil in a healthy state.

Types of Soil Conservation
1 Contour Ploughing
The furrows in which the crops are planted follow the contours. If they go up and down the slopes

they promote gullying. The photograph on the right is an aerial view of contour ploughing in Texas (U.S.A.).

2 Terracing

The slope is cut into a series of wide steps on which the crops are grown. This is very common in Asian countries in regions of rice cultivation. The rice terraces are flooded during the growing season, the water passing from one terrace to the next one below it. The flooding of the terraces is a good indication of the ability of the terraces to prevent soil erosion.

3 Planting of Shelter Belts

Belts of trees are often planted across a flat region which is liable to suffer from wind erosion. The trees break the force of the wind and thus protect the strips of land between the belts from being eroded.

4 *Strip cultivation and crop rotation*

Other farming methods include the cultivation of alternate strips at right angles to the prevailing winds, so that when one strip is laid bare for ploughing, the adjacent strip is under grass or is growing a crop. If the wind blows soil off the bare strip it will be caught and anchored by the vegetated strip. Crop rotation and the use of fertilisers ensures

Contour Ploughing in Texas

that the soil remains fertile and does not lose its 'structure'. It will therefore stick together better and be less likely to blow away.

Shelter Belts in the Dry Steppe in the U.S.S.R. The belts are of trees.

The photograph above shows gullied land in North Carolina (U.S.A.). The bottom photograph shows the same area, 20 years after it was re-afforested.

The photograph above shows gullied and eroded land in Nebraska (U.S.A.). The bottom photograph shows the same area after soil conservation methods have been used. The land is now used for growing crops.

EXERCISES

1 Briefly describe the composition and formation of soil and name the essential ingredients of a good soil.
2 Make a simple classification of soils and show how climate influences the development of soil types.
3 Briefly explain what is meant by (i) soil erosion; (ii) soil conservation. Illustrate your answer with diagrams and examples.
4 (a) Briefly explain what is meant by the terms 'soil erosion' and 'soil conservation'.
  (b) Name *three* ways in which soil erosion can be produced.
  (c) Locate and name *two* regions, one in the tropics and one in temperate latitudes, where soil erosion has actively taken place, and suggest what steps have to be taken to stop it.

Objective Exercises

1 Which of the following factors does **not** contribute to the formation of soil?
   A air and water
   B decaying plants
   C rocks and temperature
   D water-logging
   E bacteria

   A B C D E
   ☐ ☐ ☐ ☐ ☐

2 In cold wet regions leaching helps to produce grey soils which are called
   A laterites
   B chernozems
   C podzols
   D regur
   E loam

   A B C D E
   ☐ ☐ ☐ ☐ ☐

3 Alluvial soils are usually very fertile. This is because
   A they are derived only from igneous rocks
   B they are acidic
   C they consist of fine particles derived from several types of rocks
   D they often form level plains
   E loess often forms along the edges of semi-arid regions

   A B C D E
   ☐ ☐ ☐ ☐ ☐

4 In which country did wind erosion remove the top soil from a very large area, thereby creating the Dust Bowl?
   A the U.S.A.
   B Mexico
   C Canada
   D Brazil
   E Australia

   A B C D E
   ☐ ☐ ☐ ☐ ☐

5 Soil erosion can be caused in several ways. Which of the following does not cause soil erosion?
   A deflation
   B overgrazing
   C deforestation
   D weathering
   E over-cropping

   A B C D E
   ☐ ☐ ☐ ☐ ☐

6 All of the following, **except** one, account for the low fertility of soils in hot climates. Which is the exception?
   A they are leached
   B they are easily exhausted
   C they are acidic
   D they are poorly drained

   A B C D
   ☐ ☐ ☐ ☐

7 Soil erosion often results from
   A the process of weathering
   B contour strip cultivation
   C afforestation
   D deflation

   A B C D
   ☐ ☐ ☐ ☐

8 Soil erosion in a humid tropical region can be checked and corrected by
   A terracing
   B cutting down the forests
   C burning the grasslands
   D not growing crops

   A B C D
   ☐ ☐ ☐ ☐

9 Which one of the following statements is **not** true with reference to soils?
   A A laterite can form from any type of rock.
   B Hardpan is a layer of hard deposits which occurs in the lower layers of some soils.
   C Peat soils develop best under hot arid climatic conditions.
   D Terra Rossa soils form under semi-arid conditions.

   A B C D
   ☐ ☐ ☐ ☐

10 Soil erosion by rain wash can be caused by all of the following **except**
   A scanty vegetation cover
   B steep slopes
   C aridity
   D tropical rainstorms

   A B C D
   ☐ ☐ ☐ ☐

11 Soil erosion may not be prevented by
   A terracing
   B planting trees
   C contour ploughing
   D ploughing grasslands

   A B C D
   ☐ ☐ ☐ ☐

GENERAL EXERCISES

1 Explain, with the help of sketch-maps or diagrams, the meanings of three of the following statements:
   (a) Days are shorter in Peking than in Singapore during December.
   (b) Great Circles are frequently used in navigation.
   (c) The length in miles of 1° of longitude increases from latitude 80°N to the Equator.
   (d) When flying from Bombay to Manila, passengers are told to put their watches forward.

2 With the help of well-annotated diagrams, explain why:
   (a) The U.S.S.R. has a large number of time zones.
   (b) The sun never rises above the horizon at the South Pole during June.
   (c) The average temperatures are lower in regions near the Poles than in regions near the Equator.

3 Briefly explain the causes of the winter and summer seasons. Illustrate your answer with diagrams.

4 Explain the following:
   (a) The number of hours of daylight equals the number of hours of darkness along the Equator throughout the year.
   (b) Day equals night all over the earth's surface twice a year.
   (c) The sun is visible in the sky for several weeks during June at the North Pole.
   (d) Winds and ocean currents are deflected to the left in the Southern Hemisphere.

5 Write brief notes on *three* of the following, and illustrate with diagrams where appropriate: sedimentary rocks, continental shelf, earthquakes, joints, igneous rocks.

6 Choose *two* of the following features: fold mountain, rift valley, lava plateau, block mountain. For each one you choose:
   (a) Describe its appearance and explain its formation with the help of diagrams.
   (b) Name an area where an actual example could be seen.

7 (a) Draw a large, annotated diagram to show the structure of a typical volcano.
   (b) Describe how a volcano has been formed and some of the ways in which volcanic eruptions have influenced Man's activities.
   (c) Name one region which has active volcanoes.

8 Carefully explain the following processes which help in the formation of physical features, and describe the results of these:
   (a) Chemical weathering of rocks in an equatorial climate.
   (b) Mechanical weathering of rocks in a cold climate.
   (c) Weathering of limestones.

9 Write short notes on *three* of the following, and illustrate your answer with clear diagrams: artesian basin, water-table, spring, well, landslide.

10 Describe *three* of the following terms which are as-sociated with a river's course: waterfall, delta, gorge, meander. Your answer must contain relevant diagrams or sketch-maps.

11 Choose *two* of the following valley features: ox-bow lake, flood plain, rapids, terrace. For each one you choose:
   (a) With the aid of diagrams, (i) describe its appearance, and (ii) explain its formation.
   (b) Locate an area where an example could be seen.

12 The following features sometimes occur in a river's course: delta, flood plain, gorge, waterfall.
   Select *two* of these and for each one:
   (a) Locate an example by means of a sketch-map.
   (b) Describe the example and explain its formation.

13 Draw two contour maps, one of a young river valley and the other of a mature valley, and then describe the characteristic features of each. You should illustrate your answer with diagrams.

14 With the aid of clearly labelled diagrams, describe *three* of the following: spit, delta, hanging valley, caldera.

15 Illustrating your answer with diagrams and sketch-maps, describe:
   (a) The appearance and formation of a flood plain,
   (b) The importance of flood plains to agriculture in a specific Asian country.

16 You are required to make a geographical study of a small river in your country. Describe the appearance of features which you would examine and draw diagrams to illustrate these. You should give specific examples of such features from your knowledge of an actual region.

17 Name *two* features produced by wind erosion and *one* produced by wind deposition, which frequently occur in tropical deserts. For each, describe: (a) its appearance, and (b) its formation. Illustrate your answer with diagrams.

18 Choose *three* of the following rift valley, wadi, mature river valley, canyon.
   For each one you choose: (a) explain how it has been formed, and (b) give an example.

19 With the aid of well-labelled diagrams or sketch-maps:
   (a) Describe the function,
   (b) Name an actual example of *three* of the following: crater lake, delta, U-shaped valley, coral reef, loess plain.

20 Choose any *three* of the following features: ria, stack, lagoon, spit, sand-dunes, fiord. For each *one* chosen:
   (a) Describe its physical appearance.
   (b) Describe how it has been formed.
   (c) Name a region where an actual example could be seen.

21 Explain, with the aid of well-labelled diagrams, why:
   (a) Rias often provide better harbours than fiords.
   (b) Headlands have cliffs but bays have beaches.
   (c) Spits sometimes form across the mouth of a river.

22 Illustrating your answer with diagrams, describe three

ways in which mountains are formed.

23 With the aid of sketch-maps, locate examples of *two* of the following:
(a) Coral reef,
(b) A ria coast,
(c) A raised coast.
For each one you choose, describe its chief features and explain how they have been formed.

24 Name four physical features which have been formed by glacial action and for each locate a region where it could be seen. Choose *three* of these and for each:
(a) Draw a well-labelled diagram to show its chief characteristics.
(b) Describe how it has been formed.

25 In what ways does the appearance of a glaciated valley differ from that of a river valley? Draw a large contour map for each type of valley and for each, name one region where an example could be seen.

26 With the aid of annotated diagrams or sketch-maps, describe:
(a) The appearance.
(b) The function of *three* characteristic features of a glaciated highland.

27 Write a short essay on the ways in which a glaciated region can be of value to Man. Illustrate your answer with reference to specific regions where the utilisation of glacial features has already taken place.

28 Draw a large map of Africa and on it:
(i) Draw and name the Equator.
(ii) Shade one region which contains rift valleys and print on it the letters R.V.
(iii) Mark and name two important ocean currents, indicating clearly the direction in which they flow and whether they are warm or cold.
(iv) Mark (by arrows), and name, *one* wind which brings rain to South Africa and *one* wind which brings rain to North Africa. Alongside the arrows write the months during which they blow.
(v) Draw a small circle where the sun will be overhead on December 21st.
(iv) Indicate, by printing the capital letter, where you would expect to find:
(a) Basin of inland drainage (capital B)
(b) Young fold mountains (capital Y)
(c) Delta (capital D).

29 Name the fishing grounds of the world, and for any one of them describe its chief conditions which account for its abundance of fish.

30 Concisely explain why some ocean currents are warm and others are cold. Name two warm currents of the North Atlantic and two cold currents of the North Pacific. For each:
(a) State the time of year when it is most pronounced.
(b) Describe any effect which it has on the climate of a specific region.

31 Illustrating your answer with sketch-maps or diagrams, describe how a lake can be formed (a) *by erosion*, and (b) *by deposition*. For each, name a region where an actual example could be seen.

32 For any *two* of the following features: lagoon, mountain lake, delta, artesian basin, (a) describe how it has been formed, (b) make a list of its possible uses, (c) name a particular region where an example could be found.

33 (a) (i) Name *two* temperature readings which are taken at weather stations.
(ii) Name and describe the instruments used for these measurements.
(b) Clearly state what you understand by the terms: humidity, precipitation, atmospheric pressure.

34 All weather stations have a Stevenson Screen. Make a large drawing of this and then:
(a) (i) Describe its structure.
(ii) Name the instruments it contains.
(iii) Briefly describe the nature of its location.
(b) Name any two other instruments in a weather station.

35 Carefully explain the meanings of the following terms: mean daily temperature, diurnal temperature range, relative humidity.
(a) Name the instruments with which each is associated.
(b) Choose one of these instruments and briefly describe how it works.

36 A weather station contains the following instruments: wind vane, aneroid barometer, maximum and minimum thermometer. Choose two of these, and for each:
(a) describe its appearance, (b) explain how it is used to obtain weather records.

37 With the aid of diagrams explain the effect of:
(a) longitude on temperature, (b) longitude on time, (c) earth rotation on wind direction.

38 (a) Briefly describe three ways in which rainfall may be caused.
(b) Name the instrument which is used for measuring rainfall and explain how rainfall is shown on a distribution map.
(c) Name (i) a region where rain falls all the year, (ii) a region where rain falls from June to September only, (iii) a region where it rarely falls.

39 Using well-labelled diagrams, explain the meaning of *three* of the following geographical terms:
Convection Rainfall, Land and Sea Breezes, International Date Line, Tropical Cyclone, Midnight Sun.

40 (a) Draw an outline map of South America and on it shade and name (i) *one* region having an Equatorial Climate, (ii) *one* region having a Mediterranean Climate, (iii) *one* region having a Savana Climate.
(b) Describe briefly each of these climates.

41 With the aid of diagrams, explain *three* of the following: tropical cyclones, monsoon winds, prevailing winds, rain shadow. Give an actual example for each one you choose.

42 Draw a simple sketch-map of either western North America or western Europe and on it shade *three* regions, each of which has a different type of climate. Name the climatic types and then describe *two* of them.

43 Describe and account for the main features of the climates of *two* of the following regions: Burma; Northern China; the Amazon Basin. Illustrate your answer with sketch-maps or diagrams.

44 (a) (i) Name *four* types of natural vegetation which occur in Africa, and (ii) briefly describe the features of each type.
   (b) Choose *two* of the four types you have named, and for each, explain how the features you have described show the influence of climate.

45 With the aid of sketch-maps locate examples of *two* of the following:
   (a) A tropical grassland
   (b) An evergreen tropical forest
   (c) A deciduous temperate forest
   (d) A temperate grassland.
   For each one you choose, describe the characteristic features of the vegetation and show how they are related to the climate of the region.

46 Select *three* of the following: laterite, loess, lowland alluvium, desert sands, boulder-clay. For each: (a) locate and name an actual example, (b) briefly describe how it has been formed.

HINTS ON ANSWERING EXAMINATION QUESTIONS

At School Certificate level, students are expected to illustrate their answers with relevant sketch-maps and diagrams. When students are ready to take the School Certificate they should be able to draw simple diagrams of the principal landforms discussed in their physical geography course, and they should also be able to give specific examples of the more important of these. Examiners are always impressed with reference to features or processes that the students have studied at first-hand in the field.

All diagrams and sketch-maps should be bold, large and clear in outline. They should contain no irrelevant detail and any writing on the map should be in printed lettering. Coloured pencils, used in moderation, can improve a diagram and can help to make the more important pieces of information stand out clearly. All diagrams should be given a title and, where necessary, a key. The latter must always be kept as simple as possible. A common tendency of many students is to include too much information in their diagrams with the result that they become overcrowded and difficult to interpret.

Before answering any questions in the examination a student should:

1 Read the instructions at the head of the paper very carefully.
2 Read through all the questions in the relevant sections and mark those which he decides to answer.
3 Answer, first of all, the compulsory question(s) which usually carry more marks, and more time should be allocated to them.
4 Calculate how much time should be allowed for each question, making the necessary allowance for compulsory question(s). When the time limit expires, work on that answer should stop and the next question should be tackled. Any time which is left over at the end of the examination can then be used for completing unfinished answers.
5 Make sure that all answers are to the point and do not include information which has no direct bearing on the questions.

A study of past examination questions will show that a definite terminology is often used by examiners. Words and phrases in common use are *development, locate, factor, significant*, etc. The student must know the precise meaning of each of these.

EXAMINATION QUESTIONS

1 Study the climatic maps of North America in your atlas and the following figures, then:
   (a) name the state or province and the climatic region in which each place lies;
   (b) explain the pattern of temperatures and rainfalls revealed by the figures;
   (c) write a note on the natural vegetation to be seen round these places.

   (Southern Universities Board G.C.E. 1970)

2 **Either**
   (a) For **each** of **three** of the following draw a simple contoured sketch-map, numbering the contours: a glaciated mountain valley, a ria coastline, a dissected plateau, a volcano.
   (b) For any **one** of those chosen in (a), (i) describe the feature, and (ii) illustrating your answer with diagrams, suggest how it may have been formed.

   Or
   Headland and bay, anticline and syncline, rift valley and horst (block mountain), dyke and sill.
   (a) For each of **two** of the above pairs of features:
      (i) draw a labelled diagram to show the rock formations,
      (ii) name and locate an example of **each** feature.
   (b) For any **one** pair you have chosen in (a), suggest how **each** feature may have been formed.

   (Cambridge G.C.E. 1970)

3 Explain why:
   (a) Evergreen woodland and scrub occur in areas of hot dry summers and warm wet winters.
   (b) Grassland vegetation occurs in areas of hot moist summers and very cold dry winters.
   (c) Very dense evergreen forest occurs in areas of very hot and very wet conditions throughout the year.

   (Cambridge G.C.E. 1970)

4 With the aid of diagrams, describe and give reasons for **three** of the following:
   (a) Daily land and sea breezes.
   (b) Chinook (föhn) winds.
   (c) The Trade Winds.
   (d) The frequent changes in the direction of the wind over Britain.

   (Cambridge G.C.E. 1970)

5 **Either** Discuss the causes and effects of river capture.
   Or Explain the development of superimposed drainage. Illustrate your answer with sketch-maps of **named** examples.

   (Oxford/Cambridge Alt 'O' 1971)

6 **Either** Illustrating your answer with a sketch map, briefly describe and account for major variations in climate in North America or Africa.
   Or Describe and account for the monsoon climate of the Indian subcontinent.

   (Oxford/Cambridge Alt 'O' 1971)

7 Quoting examples and illustrating your answer with diagrams explain how **two** of the following physical features are formed:
   (a) canyons,
   (b) rift valleys,
   (c) fiords,
   (d) deltas

   (Southern Universities Board G.C.E. 1971)

8 (a) Outline briefly the principal causes of the surface currents of the oceans.
   (b) Draw a sketch-map to show the arrangement of the ocean currents north of the equator in **either** the Atlantic **or** the Pacific. Mark and name **two** warm and **two** cold currents.
   (c) Explain with examples the part played by ocean current in
      (i) the formation of fog, and
      (ii) the development of fishing grounds.

   (Cambridge G.C.E. 1971)

9 **Either,** (a) (i) Describe and account for **three** of the landforms associated with hot deserts.
   (ii) Show how physical conditions and human activities can lead to the establishment of settlements even in the hot deserts.
   or, (b) (i) Describe and account for the landforms associated with the floodplains of large rivers.
   (ii) Where are human settlements most likely to be found in floodplain areas, and why?

   (Oxford G.C.E. 1972)

10 How are waves formed in the sea? How does wave and current action influence the form of coastlines.

   (Oxford/Cambridge Alt 'O' 1972)

11 **Either** Describe, mentioning **named** examples, the ways in which natural lakes may be formed.
   Or Write a short essay on springs and underground water.

   (Oxford/Cambridge Alt 'O' 1972)

12 **Either** What features characterise temperate maritime and temperate interior climates? Why do they differ? **Or** Briefly describe and account for the principal regional variations of climate in Europe.

(Oxford/Cambridge Alt 'O' 1972)

13 (a) Name the instruments used in a school weather station to record temperature, humidity, and rainfall.
  (b) With the aid of diagrams, describe any **two** of these.
  (c) Explain how the following are calculated:
    (i)   the daily average temperature;
    (ii) the monthly average temperature;
    (iii) the monthly average rainfall.

(Southern Universities Board G.C.E. 1972)

14 (i)   Explain, with the aid of diagrams, what is meant by latitude and longitude.
  (ii) Calculate the time at Greenwich when it is 10.00 p.m. in Buenos Aires (Longitude 60°W)
  (iii) Explain why at the Poles the sun does not set for many weeks on end at certain times of the year.

(Welsh Joint Education Committee G.E.C. 1972)

15 With the aid of diagrams explain fully the nature of and the reasons for.
  (i)   'Chinook' winds in Canada
  (ii) The general decrease in temperature polewards with increasing latitude from the equator.
  (iii) The wide diurnal variations in temperature that occur in Hot Deserts

(Welsh Joint Education Committee G.E.C. 1972)

16 (a) Outline the chief factors which affect the amount and nature of the weathering of rocks.
  (b) Describe, with the aid of diagrams, the formation of three of the following: *rounded boulders of hot desert areas, screes, earth pillars, clints and grykes.*

(Cambridge G.C.E. 1972)

17 (a) With reference to **one** type of forest:
    (i) draw a sketch map to locate an extensive area where it is found,
    (ii) show how the vegetation is adapted to the climatic conditions.
  (b) Describe, with examples, how man has influenced the world's natural vegetation cover.

(Cambridge G.C.E. 1972)

18 (a) With the aid of diagrams explain why it is nearly midday in New Zealand when it is midnight in England.
  (b) When it is midnight in England, what is the time in Canberra, Australia?
  (c) With the aid of diagrams explain why all areas north of the Arctic Circle have 24 hours daylight on 21st June, and 12 hours on 21st March.
  (d) For 21st December give the number of hours of daylight at
    (i)   the Arctic Circle
    (ii) the Equator
    (iii) the Antarctic Circle.

(Southern Universities Board G.C.E. 1972)

19 EITHER: State the landforms associated with two contrasting rock types. In your account give clearly the reasons for the similarities and for the differences.
  OR: (a) Draw and label a diagram of a composite volcano.
      (b) Describe
          (i) the eruption of a explosive volcano and
          (ii) the formation of sills and dykes.
      (c) Explain why volcanoes and earthquakes often occur in fold mountain areas.

(Southern Universities Board G.C.E. 1975)

20

27°5'N		2°30'E			SOUTHERN ALGERIA						
	J	F	M	A	M	J	J	A	S	O	N	D
°C	13	16	20	25	29	35	37	36	33	27	19	14
rain (mm)	2·5	1·3	1·3	1·1	1·1	0	0	1·1	1·2	1·4	2·5	5·1

Mean daily maximum and minimum temperature	
45°C	28°C July
21°C	6°C January

  (a) Give reasons to explain
    (i)   the temperature figures
    (ii) the low rainfall and the pattern of distribution throughout the year and
    (iii) the differences between the maximum and minimum temperatures in July and January
  (b) Describe and account for
  EITHER (i) the physical landforms
  OR   (ii) the vegetation of a tropical desert area.

(Southern Universities Board G.C.E. 1975)

21 (a) Illustrating your answer with diagrams explain why the length of day and night varies at different latitudes and seasons.

(b) Explain why the Equator is the longest line of latitude.

(c) What are Great Circles and why are they important in navigation?

(Southern Universities Board G.C.E. 1975)

22 Describe the principal types of volcanic landform and show how their shape is related to the type of material of which the landform is composed.

(Oxford/Cambridge Alt. 'O' 1975)

23 Describe and indicate the mode of formation of the main features which may have been produced by ice as an agent of erosion in highland Britain.

(Oxford/Cambridge Alt. 'O' 1975)

24 What conditions favour the growth of coral? Outline some of the theories advanced to account for the formation of fringing reefs, barrier reefs and atolls.

(Oxford/Cambridge Alt. 'O' 1975)

25 What factors influence the distribution of rainfall over the earth's surface?

(Oxford/Cambridge Alt. 'O' 1975)

26 Briefly describe and explain the weather associated with
(a) a cold front,
(b) a winter anticyclone.

(Oxford/Cambridge Alt. 'O' 1975)

27 (a) Briefly distinguish between weathering and erosion.
(b) For **each** of physical (mechanical) weathering and chemical weathering:
(i) Name and describe a feature formed by the process.
(ii) Explain how the feature may have been formed.
(c) Explain how biological (organic) weather ng helps in the decomposition and disintegration of solid rock.

(Cambridge G.C.E. 1975)

28 Drumlins, eskers, terminal moraines and kames are all features of glacial or fluvio-glacial deposition.
(a) For each of any **three** describe its main features and the materials of which it is made.
(b) For **each** of **two** of those selected in (a) state the factors which were important in its formation and illustrate your answers by labelled diagrams.

(Cambridge G.C.E. 1975)

29 (a) (i) Describe **three** processes of erosion by which a river deepens and widens its bed.
(ii) State **three** ways by which a river transports its load.
(iii) Where and why does deposition occur in the course of a river?
(b) For **each** of the following, explain with the aid of diagrams how they may have been formed by the action of a river
(i) river bed pot holes, (ii) levées.

(Cambridge G.C.E. 1975)

30 Study the weather map on page 175.
(a) Name (i) the weather system over Scandinavia and (ii) that over Central Europe.
(b) What is the meaning of each of the following symbols?

(i) $\overset{\triangle}{\triangledown}$ (ii) R

(c) Describe the weather being experienced over Scotland.
(d) Explain why
(i) snow is falling at station A,
(ii) it is colder at station A than at station B,
(iii) there are light S.E. winds at station C,
(iv) there is fog at London (D).

(Cambridge G.C.E. 1975)

31 The following weather readings were taken at a school weather station: Pressure 1026 mb; max. temp. 15°C (59°F); visibility 1000 yards; rainfall trace.
(a) State very briefly what you understand by the term 'weather'.
(b) Explain the meaning of **each** of the readings given above.
(c) Draw a simple labelled diagram of the instrument which records changes in pressure as they occur. Name the instrument.
(d) Explain how you would read and reset the instrument used to measure maximum temperature.
(e) State **two** precautions you would take in siting the rain gauge. Give a reason for each.

(Cambridge G.C.E. 1975)

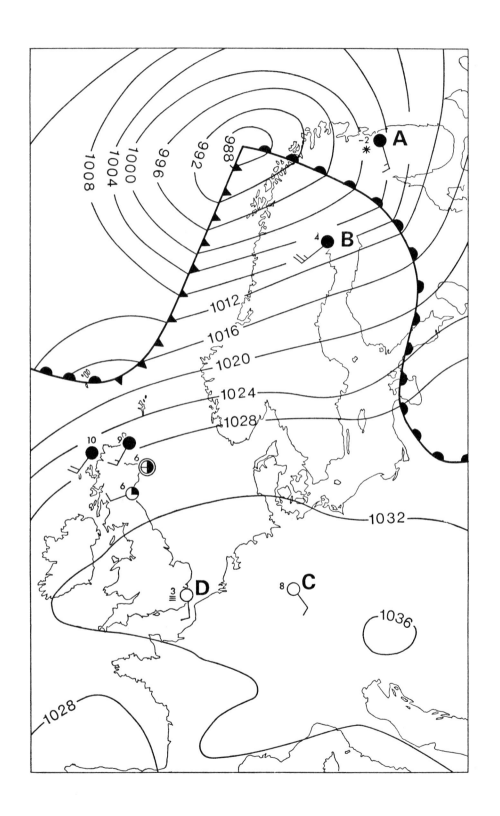

INDEX